Management Information Systems and Workflow Automation

Other Labels & Labeling books:

For the latest list please visit: **www.labelsandlabeling.com**

Management Information Systems and Workflow Automation

Michael Fairley LCG, FIP3 and FIOM3

Management Information Systems and Workflow Automation

First edition published 2017 by:

Tarsus Exhibitions & Publishing Ltd

Printed by CreateSpace, an Amazon.com company.

ISBN 978-1-910507-12-4

Contents

While every care has been taken to ensure the information, charts, diagrams and illustrations in this publication are correct at the time of publishing it is possible that technology, specifications, markets and applications, or terminology may change at any time, or that the editor's or contributor's research or interpretation may not be regarded as the latest accepted guidance in some parts of the world of labels.

The publishers therefore cannot accept responsibility for any errors of interpretation or for any actions, decisions or practices that readers may take based on the publication content and would advise that the latest industry supplier specifications, standards, legislative requirements, performance guidelines, practices and methodology should always be sought before any investment or implementation is made.

Foreword

This book clearly and intelligently illustrates the requirement for the modern forward thinking label printing company to be in accurate control of all aspects of its day to day sales and manufacturing business. In today's fast paced global economy sales opportunities are won and lost in the blink of an eye, brand and retailer competition is so fierce that speeds to market have moved to shorter and more impossible deadlines and ever present focus on budget costs.

These pressures can exert huge strains both financially and operationally on the modern label printer. In today's business environment every member of a company's operation must have clear, concise and most importantly accurate information to enable that individual to play their part in successfully navigating a client's job/brand through the many stages and processes of the typical job lifecycle.

Successfully within this publication Mike Fairley has focused hard on the joined up approach to electronically managing an efficient twenty first century operation, placing emphasis on fast accurate estimating, procurement, planning, asset data collection and post data analysis. This book gives insight into the functionality of the fully integrated management information system providing operators and senior managers with insight of how the real time information can assist in the long term strategic decision making, and the day to day problem solving that many of us can identify with in our own businesses.

I believe the contents of this book will not just prove invaluable as a tool to the growing company making its first moves into the complex world of Management Information System integration. Whilst experienced users will find it useful as a benchmarking guide, to perhaps uncover the potential possibilities and opportunities that might lay either untapped or under- utilised within their own MIS systems, already embedded in their operations.

"If we can't measure it, we can't report on it, if we can't report it then we cannot control it" and without control we are not masters of our own destiny.

Christopher J. Ellison
Managing Director, OPM (Labels & Packaging) Group Limited
Vice President, Finat

Preface

The label and package printing industries have undergone a quite significant change in recent years, with run lengths decreasing, more product versions and variations, increasing personalization, the need to track and trace products, the use of social media, together with technology advances that include digital printing, combination and hybrid presses, more advanced inspection and finishing, increasingly sophisticated pre-press, quality control, color management and production automation, as well as a whole range of environmental and sustainability issues.

Such changes have meant that the management of a label or package printing plant has become increasingly complex and time consuming. More complexity, more shorter run jobs, multiple versions to process and produce, will almost certainly put pressure on people, time and administration – from estimating, order processing, job management, production scheduling, inventory control, quality control, costing, invoicing and shipping through to the overall accounting and financial management of the business. If not managed well all kinds of bottlenecks will occur, jobs get delayed, quality gets compromised, and the company's margins and profitability soon start to suffer.

To streamline this administration process, manage and control an ever more automated production facility and have instant access to all the job, production, and financial facts and figures that can enhance performance and profitability, a number of industry-dedicated Management Information Systems (MIS), specialized and niche software solutions, and evolving hardware have been developed and introduced. These are certainly now having a positive impact on the enhanced profitability of many of the industry's MIS adopters. Indeed, it almost seems impossible for label and package printing businesses of tomorrow to remain competitive and profitable without such systems.

Add to these MIS advances the increasingly sophisticated automation of both digital and conventional presses, 100% inspection of webs and bar code verification, dedicated color management and control systems, automated set up of slitting knives, laser die-cutting machines, varnishing and foiling units, and it can soon be realized that the label and package printing industries are rapidly moving towards the day when press rooms will become streamlined, hands-free, 24 hours a day operations. Maybe still a few years away, but almost certainly will become a reality.

The aim of this book therefore is to explore and describe what MIS and workflow automation systems are able to offer – both today and for the future – and to provide guidance on choosing and implementing successful and suitable procedures, systems and technology. A handy Appendix of Suppliers is also included. Hopefully the book will provide a valuable and convenient reference source for label converters and package printers, industry suppliers and perhaps label end users that buy and specify labels.

Michael Fairley
Director, Labels & Labelling Consultancy
Founder, Label Academy

About the Label Academy

This book is part of the recommended study material for the Label Academy, a global training and certification program for the label industry. The Label Academy was created by the team behind Labels & Labeling magazine and the Labelexpo series of events.

The Academy consists of a series of self-study modules, combining free access to relevant articles and videos with paid text books (both printed and electronic). Once a student has completed a module, there is an opportunity to take an online test and earn a certificate.

It is expected that a Label Academy qualification will become a standard in the industry – for printers/converters, suppliers, brand owners and designers – and assist in providing a benchmark. In addition to its own training, the Label Academy will aim to become a resource provider to the many existing educational programs in the industry. Accredited training courses will be promoted through the Label Academy website and books will be provided at discounted rates.

The Label Academy concept was pioneered by industry expert Mike Fairley. This was in response to a reduction in the number of dedicated printing colleges and the need to standardize training across the world. The label industry also has its own specific training needs – it has some of the widest range of materials, printing processes and finishing solutions of any printing sector.

We are also working with other training experts and authors to ensure that the Label Academy provides up-to-date and relevant training material for the industry.

The Label Academy is supported by the key trade associations, including FINAT, TLMI and the LMAI.

www.label-academy.com

Label Academy sponsors

Thank you to our founding sponsors, without whom this ambitious project would not have been possible:

Cerm
Cerm designs business automation software solutions to meet the specific demands of flexo and digital narrow web printers. Using the latest technology, our team's focus is on innovation and continuous improvement.

Our automation solutions support each step in the printer's integrated workflow – from estimating to production, shipment and data collection – and provide the feature and functionality printers need to gain efficiency and improve profitability.

Cerm inspires collaboration and helps printers remain competitive in the market and deliver the best products possible. We are proud to sponsor the Label Academy and contribute to the future of the narrow web printing industry.
www.cerm.net

Flint Group Narrow Web
Flint Group Narrow Web has the products, the solutions, and the technical experts to handle any print situation. Providing solutions for food packaging, sustainability, increased bottom line, efficiency, and uptime – delivering the basics needed to run a successful operation, and the expertise to go above and beyond to another level of success.

Our experts provide solutions to your printing problems with the innovative products and services that have made us an industry leader around the world. Wherever you are, we are – available to help you reach your business goals today and into the future.

Continuous improvement is paramount to Flint Group; we are proud to sponsor the Label Academy and the benefits it will bring to the future of our industry.
www.flintgrp.com

Gallus Group

The Gallus Group with its production sites in Switzerland and Germany is a leader in the development, production and sale of narrow-web, reel-fed presses designed for label manufacturers. The machine portfolio is augmented by a broad range of screen printing plates (Gallus Screeny), globally decentralized service operations, and a broad offering of printing accessories and replacement parts. The comprehensive portfolio also includes consulting services provided by label experts in all relevant printing and process engineering tasks. The Gallus Group is a member of the Heidelberg Group and employs around 430 people, of whom 253 are based in Switzerland. The group headquarters is in St.Gallen, Switzerland.

www.gallus-group.com

MPS Systems B.V.

Producing high-quality label printing depends on several factors; one of them is the operator of the press.

As a press machine builder since 1996, MPS Systems B.V. knows how important training and education on subjects like pre-press, label printing and finishing is. For label printers, it is critical that their operators keep up with pre-press and press developments in addition to label trends. Therefore, MPS sponsors the Label Academy, to advance operator's passion for printing, share expertise and help multiply benefits.

The MPS slogans of 'Printers First' and 'Technology with Respect' have always underlined the core philosophy of MPS from press design to operator satisfaction. We develop our presses with a strong focus on user-friendliness and respect for the press operator: Printers First.

www.mps4u.com

HP Indigo

HP Indigo is a global leader in digital printing, with a broad portfolio of digital presses and workflow solutions. Indigo's proprietary Liquid Electrophotography (LEP) technology delivers exceptional print quality for the widest variety of applications including labels, flexible packaging, shrink sleeves and folding cartons. HP Indigo's digital presses match gravure print quality satisfying the most demanding brands.

A division of HP Inc.'s Graphics Solutions Business, Indigo serves customers in more than 122 countries, including many of the top label and packaging converters worldwide.

www.hp.com/go/labelsandpackaging

UPM Raflatac

In a little more than three decades, UPM Raflatac has become one of the world's leading manufacturers of pressure sensitive label materials, developing and leveraging the latest innovations in adhesive technology. Our film and paper label stocks are used for product and information labeling across a wide range of end-uses – from pharmaceuticals and security to food and beverage applications.

We are an engineering driven company with industry-leading products known for their consistent high quality and top performance. We are also known for the high performing supply chain and undisputed leadership in the area of sustainability. UPM Raflatac's dedication to innovation, sustainability and top quality is matched only by our commitment to service excellence. We call it the Raflatouch.

www.upmraflatac.com

About the author

Michael Fairley
Director, Labels & Labelling Consultancy
Founder, Label Academy

Michael Fairley has been writing and speaking about label and packaging materials, technology and applications since the 1970s, both as the founder of Labels & Labeling and other print industry magazine titles and as an international consultant writing or contributing to label industry market and technology research reports for the likes of Frost & Sullivan, Economist Intelligence Unit, Pira, InfoTrends and Labels & Labelling Consultancy.

He is the author of the Encylopedia of Label Technology, co-author of the Encylopedia of Brand Protection, a contributing author to the Encylopedia of Packaging Technology and a contributing author to the Encylopedia of Occupational Health and Safety. He also provided significant input to the Academic American Encylopedia.

He now works as a consultant to Tarsus Exhibitions & Publishing – which organizes the Labelexpo shows, Label Summits and publishes Labels & Labeling magazine – as well as regularly speaking at industry conferences and seminars.

He is a Fellow of the Institute of Packaging / Packaging Society, Fellow of IP3 (formerly the Institute of Printing), a Freeman of the Worshipful Company of Stationers, an Honorary Life Member of FINAT and a Licentiate of the City & Guilds of London Institute. He was awarded the R. Stanton Avery Lifetime Achievement Award in 2009.

Acknowledgements

Writing a handbook on the nature and use of Management Information Systems and Workflow Automation has been something of a challenge. Much of what is now achievable with such systems has only become possible relatively recently, and continues to change quite rapidly. Integration of MIS with both digital and conventional production technology – including presses, inspection, verification and finishing – is evolving all the time, while moves towards full production automation continue apace.

Consequently, the compiling and preparation of this handbook has involved extensive research and study, reviewing MIS supplier literature and websites, talking with vendors and users of systems, looking at software and technology solutions at Drupa and Labelexpo shows, as well as considerable desk research.

Certainly, there is a lot of information available to label and package printers, but collecting and analysing it before making any decisions could take a considerable amount of time. This book has therefore been designed to do much of that homework for the industry, bringing together a collective summarization of what's possible and what's available – and even trying to look ahead to the future where production of labels and packaging may be virtually hands-free.

Extensive thanks are therefore due to all the supplier companies that have directly or indirectly supplied material for the book; in particular ABG International, Adents, AVT, Bobst, Cerm, Comexi, CRC Information Systems, EFI, Erhardt + Leimer, Esko, Fusion Technology, Gallus, Globe-Tek, GMG, HP Indigo, Imprint-MIS, Label Traxx, MPS, Nilpeter, OneVision, Optimus, PC Industries, PrintMIS, QuadTech, Tharstern, Sistrade, Xeikon, and X-Rite.

Particular thanks are also due to Cerm, Label Traxx and Tharstern for all their help in providing multiple screen-grab illustrations used throughout the book, and to Geert van Damme (Cerm) and Katy Nightingale (Label Traxx) for giving up some of their valuable time to read through chapters and provide additional content and feedback. The help, support and encouragement of these companies and individuals has been invaluable.

Chapter 1

Managing information in label and package printing

When computers first started to be used in the label and package printing industries in the early 1980s, dedicated software was written for label estimating and costing, later expanding its capabilities into invoicing, shipping data and label generation.

More complex management information software was developed for commercial printing operations, extending beyond costing and estimating into order processing and materials management and the beginnings of sophisticated database resources.

Many of these early software companies were absorbed into larger groups, which developed sophisticated systems that now cover all areas of label and package printing industry management, including estimating, order processing, job management, production scheduling, shipping and invoicing, stock and materials management, through to costing and shop floor data collection.

The latest technologies look to drive increased workflow automation, integrating with pre-press, inspection and finishing, all working to lower production costs and increase profitability, as well as providing easy-to-use E-commerce software.

These systems are generally brought together under the heading of Management Information Systems, more commonly abbreviated to MIS. 'Information' to be managed is both electronic and physical, including paper and electronic documents, audio and video information, spreadsheets and data files, all of which need to be processed and delivered through multiple channels including e-mail, smart phone, printer or web interface.

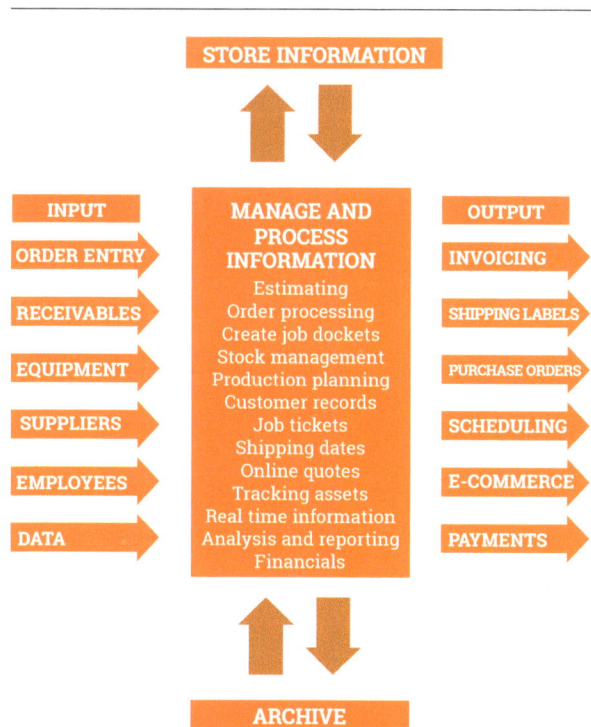

Figure 1.1 The basic structure of a computerized management information system

INPUT	PROCESSING	OUTPUT
Customer information	Estimating	Quotes and contracts
Supplier information	Management of product information and approval cycle	Sales order
Request for quote	Order processing	Confirmations
Product description	Production job creation	Works instructions:
Sales Order	Production scheduling	• Per job
Order changes	Stock/inventory movements/ changes	• Per function
Materials information	Shop floor data	• Per production
Purchase order	Tools management	Purchase orders
Confirmation	Manage/organize packing and shipping	Scheduling
Receivables	Costing/Invoicing	Identification labels
Purchases invoices	Month or job end closing	Shipping notes
Employees/ equipment		Invoice
Imports from accounts		Analysis and reporting
Data		E-information
		Export to accounts

Figure 1.2 Shows what may be included in an expanded management information system

In this book we define a management information system as an organized combination of people, hardware, software and communication networks both within an organization and linking to the outside business world. The system is designed to collect information covering all important business processes, then record, process and route that data to where in the business it is needed. It provides detailed analysis and reports which assist management in understanding the effects of different strategies, and helps in effective decision making.

Let's now apply these basic principles to a label converter or carton business. What information does it need to capture (input data)? How is it managed? What needs to be stored and/or archived, and what needs to be delivered (output data)? Figure 1.1 outlines a simplified label information management system.

In more sophisticated MIS systems the range of input, processing and output steps becomes more complex, as seen in Figure 1.2.

Following on from the information provided in

these charts, label and package printing industry MIS systems can be divided into the following stages, or categories, of operation:

- **Transaction processing:** handles all routine and recurring transactions, such as invoicing, supplier payments, payroll and inventory tracking.
- **Operational information processing:** gathering and organizing information from transactional processing and other forms of performance-related data. then presenting it in a format useful for managing the business.
- **Decision-supporting information:** provides managers with the necessary data and support information from a wide range of sources to enable them to make intelligent and informed management decisions.
- **Problem solving:** the computer uses captured information to help recognize, formulate and solve problems. It may also be able to explain solutions and, depending on the level of sophistication, learn from that problem solving.

An efficient, label specific management information system is able to manage the entire business, streamline the administration process, reduce costs, eliminate errors from the re-entry of data and maximize the efficiency of plant personnel.

Key benefits will include:
- Identify the company's strengths and weaknesses and so enable continuous improvement from a measured base
- Provide an overall picture of the company and the way it operates
- Improve decision making and speed up actions
- Better manage customer information and target sales, marketing and promotional activities
- Gain a competitive advantage

We now review in more detail the most common operations and functions of a modern label or package printing MIS and workflow automation system.

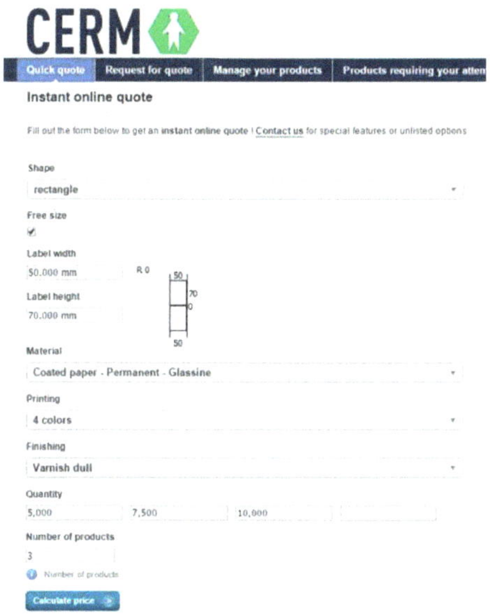

Figure 1.3 Instant on-line label quote form using Cerm MIS software

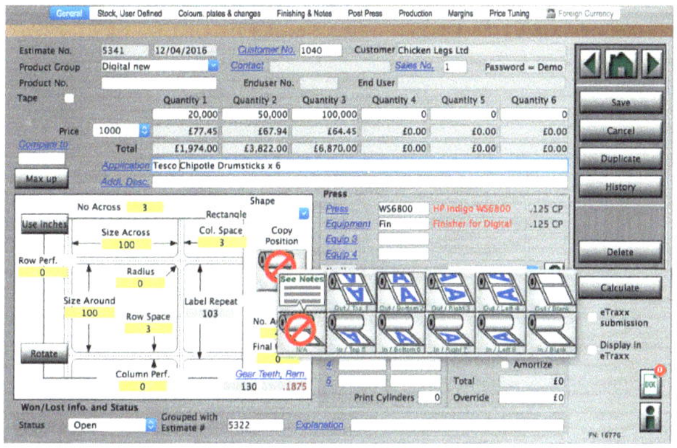

Figure 1.4 Label Traxx Online label estimate process showing items such as quantity, wind direction, size, press and price per thousand

ESTIMATING

This is an essential element of any business management system in a label or carton plant, with customers expecting to receive an estimate quickly and efficiently.

The request for a quote should describe the external characteristics of the label - for example size/shape, wind direction, number of colors, paper and quantities - and may need to include different quantities, print or finishing options. The converter will also be looking at whether printing the proposed job by conventional or digital processes offers the best return. An example of an online quote form, using Cerm software, can be seen in Figure 1.3.

After an estimator has established the cost, somebody in the company will define the proposed sales price, normally by adding a margin or evaluating added value contained in the estimate.

To prepare the estimate the converter may additionally need to search a database for appropriate substrates, inks, tooling or cylinders and determine their availability. The estimate may need further manual tuning in relation to competitor or profitability metrics before it is printed onto the company's letterheads, or sent to the customer by fax or e-mail. If the customer subsequently telephones to discuss the quote, it can be called-up on screen for review and possible amendment.

MIS software can be specified with an 'online quote' facility that will automatically calculate the selling price, or quote for a variant of an existing estimate (such as another quantity), following pre-defined rules.

Sophisticated Online quote software allows virtually any member of staff to produce an accurate quote, eliminating delays in getting back to customers. Some software will even enable customers to quote jobs themselves.

Web-based systems provide on-the-road sales staff with immediate answers to scheduling or costing questions along with detailed quotations and even facilities to upload the artwork for the job.

Another role of an estimating software module is to allow converters thinking of investing in a new press to run trial estimates and do cost crossover comparisons with their existing machinery or with other suppliers' press technology.

Figure 1.5 Viewing of the job ticket on the shop floor using Tharstern MIS software without the need to re-key any information. Re-orders can also be quickly expedited through a simple job duplication

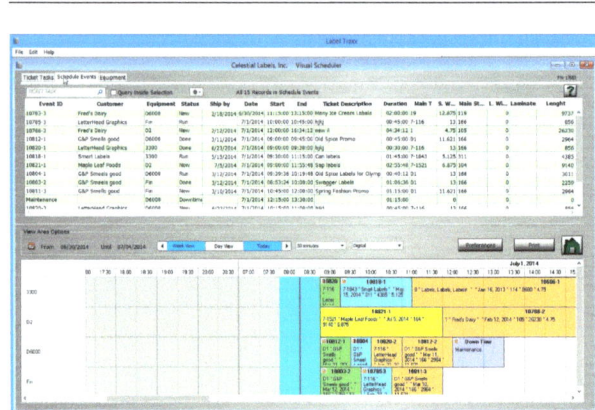

Figure 1.6 Visual scheduler in Label Traxx MIS software system

DATBASE OBJECTS

An extension of estimating software modules allows converters to identify every individual product requested by the customer and perform the following actions:

- Follow up the Artwork approval cycle (upload artwork, transfer to prepress, receive soft proof & thumbnail, receive approval/reject, …)
- Follow up any history of repeats and traceability (when did we produce, how many, with which paper, which box numbers)
- Optimize production formats – for example combine products with the same characteristics (for example same substrate or die) into a single job
- Display stock level / orders per product.

The system stores this information as 'objects' (or 'Products' in the Cerm software), to which specific customer order information is added - list of products, quantities, delivery date(s) and addresses(s) – when a specific quote is required. The software then calculates the cheapest of the pre-defined production paths and creates a production job.

A major advantage of this approach is that product approval can be managed before a sales order is entered, significantly decreasing lead-times.

ORDER PROCESSING

Once an estimate has been agreed, it will need to be efficiently and accurately turned into an order, with the job passing from sales into production and eventually on to despatch to the customer. The aim of the order processing software therefore is to accurately convert the estimate into the production order (create the job bag or job ticket. See Figure 1.5.)

With order processing information entered, production planning can start. These functions include: assign job priority; track jobs in progress; view press or process capacity; track and maintain roll materials inventory (using bar codes) and order and receive materials electronically; prepare and issue delivery notes and print-out pallet, carton or shipping labels.

PRODUCTION SCHEDULING

Production scheduling MIS modules provide full visibility of machine capacity (see Figure 1.6), both short and long term, and offer detailed scheduling of each production step from customer approval through to shipping. It may also be possible for press operators to consult their work schedule on their press or feed back what they are doing in real time. A change in schedule can be automatically seen on the shop floor. It is also possible to reserve

Figure 1.7 Barcode scanning of rolls using an iPod scanner. Source Label Traxx

production slots and automatically schedule planned maintenance stops.

Job optimization steps include grouping or ganging similar labels or cartons together by substrate, colors or cutter tooling for example. Data can be sent out to a system like the EskoArtwork Automation Engine for automated step and repeat, while some digital presses come with their own built-in step and repeat software.

CUSTOMER ACCESS

Some suppliers now offer job tracking modules that allow customers to view their own jobs and even enter information directly into the MIS system in a secure environment with multiple levels of access.

Customers can raise quotes and jobs, review progress of jobs already in hand, monitor key milestones for overall tracking, and even ask for customer specific reports to be run in real time. If finished print is held in stock, customers can raise 'Call Off' orders that automatically feed through to warehousing modules for picking, packing and delivery.

INVENTORY CONTROL AND PURCHASING

All label and packaging printing converters carry stock from paper and film reels or sheet stock, to inks and varnishes, cutting, embossing and foiling dies, cartons and product packing materials. In addition, they may be holding stocks of labels for customer call-off, or pre-printed stock products labels. And consumables like labelstocks, inks, cutters and other tooling, packing and cleaning materials, foils and lamps are coming in all the time.

Controlling and managing both stocked products and finished goods inventory is therefore a key part of any label management information system – ideally tracking product inventory in real time. It also becomes feasible to agree and store both minimum and maximum stock holdings, with the system providing replenishment reporting and even automated inventory management.

Most label-specific MIS systems have a roll tracking facility, based on bar coding, enabling full traceability of the rolls used on any job. Some are fully integrated with material suppliers, allowing electronic ordering. The shipment details are sent back to the MIS electronically, with the manufacturer's EPSMA barcode number of the roll and the exact width and length. A handheld scanner (Figure 1.7) can then be used to take roll inventory by scanning all of the roll barcodes.

Finished goods inventory includes labels called-off by customers and items that are resold like ribbons and printers. Increasingly, MIS suppliers are also being asked to incorporate interfaces to transport suppliers, enabling the sending of electronic instructions to DHL, UPS, etc., in order to manage transport and shipping.

QUALITY CONTROL

Quality control systems which document quality issues, returns procedures and complaints logging have become an important element in the management of label or package printing businesses.

When problems or complaints arise there needs to be a system of reporting and corrective actions, including the generation of management reports that enable fault trends to be ascertained and analysed.

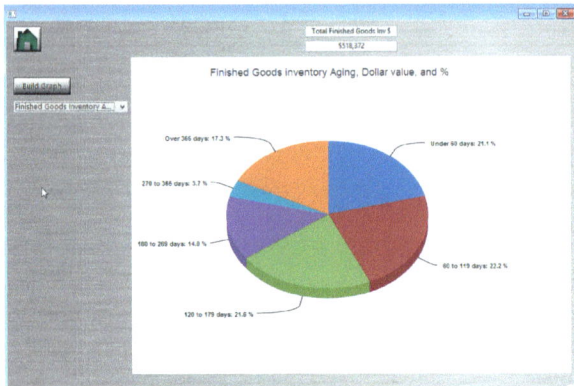

Figure 1.8 Finished goods inventory as shown in the Label Traxx system.

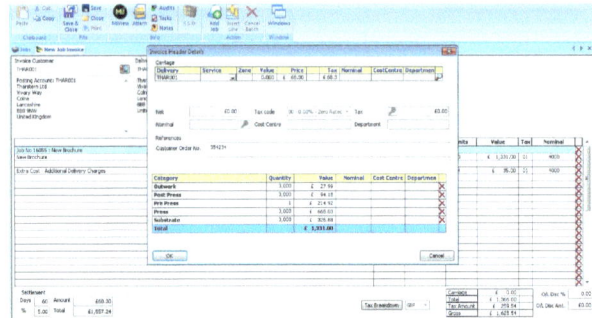

Figure 1.9 Sales and purchase invoice reporting in Tharstern MIS

This may lead in turn to updating of quality control procedures or new employee training documentation.

Quality control systems may also store ISO, OSHA or other international or national standards requirements and documentation that can be viewed or print generated by employees.

JOB COSTING

Once a job has been completed and shipped it needs to be costed and an invoice issued as quickly as possible. Additionally, the sales, accounting and management teams will want to know if they made a profit on the job and how much.

Label costing systems can now be linked to a wide range of company data, including the clocking in and out of each employee - with time being automatically recorded to the job docket. Labelstock, inks and other consumable information is automatically captured. This information is combined to create the actual job cost, which can then be compared against the original estimate.

Some variables are more difficult to measure – ink consumption for example. This is because ink is sometimes mixed out of other inks without real booking, and it is not easy to divide the total consumption of the same ink between different jobs. Interfacing with ink dispensing systems, such as

those manufactured by GSE, will ensure proper booking and traceability of ink.

For converters requiring more in depth shop floor data, the latest MIS modules provide real-time press room statistics including the number of labels or sheets produced, material used and running speeds. Data can include full capture of both gross and net label quantities as well as length or quantity converted, so providing accurate cost monitoring control throughout production.

ACCOUNTING

The Accounts operation is at the heart of any label or package printing industry management information system. It operates the general ledger/nominal ledger where all financial transactions are received, processed and summarized in real-time. The results of these transactions are shown in financial reports which provide the management team with forecasts, profit and loss data, and all the usual financial management data provided by standard ledger analysis tools.

This makes it possible to view critical business data at a glance, track the company's progress against key business performance indicators, obtain an instant picture of how the business is performing and quickly highlight any areas in need of attention. See Figure 1.8. for an example of Finished Goods inventory by age, value and percentage.

Where required, a whole variety of more

comprehensive management reports can be prepared and presented. These will enable the management team to drill down much further for more detailed analysis.

The accounts department is also involved in the invoicing process, in sales and purchase invoice reporting (Figure 1. 9), credit control and surveillance, handling tax rates, pre-payments, and dealing with accounts receivable.

MIS modules allow the accounts department to manage employee expenses, supplier invoices, cash flows, reconcile accounts and offer historical reporting. Currency handling for forex (foreign exchange) sourcing or sales transactions might also be included.

WORKFLOW AUTOMATION

The label and package printing industries are continually seeking ways to integrate MIS with increasingly automated pre-press and production systems to meet a range of challenges: to remove the risks of human error; to handle an ever-increasing number of shorter runs; and to overcome a shortage of skilled operators. In addition, the impressive rise of digital print for labels has increasingly pushed both MIS and pre-press vendors to develop more automated solutions.

To remain profitable today, converters must ensure their pre-press and production workflows are integrated within their business and management operations and connected with their entire supply chain, 24/7, wherever that may be in the world. Indeed, some systems now offer the ability to take an order online, accept payments, pre-flight files, correct them, and send directly to the press without any operator intervention.

Overall, the trend is towards ever more press and finishing line automation. We are seeing the emergence of self-managing presses, cloud computing and cloud-based assistants, smart data management and smart printing systems, WiFi control, and even fully hands-free and totally 'lights-out' production. This trend affects all aspects of printing press and finishing line technology.

Traditional narrow web machinery suppliers are seeing the press less as a mechanical device and

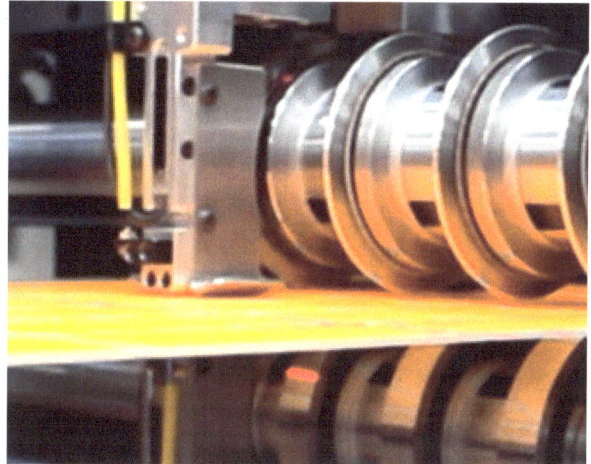

Figure 1.10 Automatic re-setting of slitting knives with workflow automation. Source ABG International

more as a software-driven machine tool, with the prepress job data file increasingly driving the whole workflow. The JDF sent to the printing press contains the job protocol, in CIP3 or similar format, which will transfer job data such as pressure accuracy, dot gain, register, web control, and cutting depth.

Integration of MIS and pre-press automation software ensures that customer 'job' information held in the estimating files is used to automatically create a new pre-press job and deliver new artwork, make it print-ready and prepare proofs. Status updates on everything from plate layout, RIP-ing and plate making are sent back to the MIS.

Whatever printing process is being used – flexo, offset, letterpress, digital – brand colors must all be accurate and consistent, and can be controlled through the same workflow.

Workflow automation can identify which jobs – typically reprints – can re-use an existing plate set, or an existing cutting die from store, or enable a slitter operator to retrieve previous slitter instructions to automatically re-set the slitting knives. See Figure 1.10.

Further integration with camera inspection systems - such as that demonstrated by AVT and

Cerm – allows the creation of inspection files per print frame (Figure 1.11). A barcode is then printed for every print frame in the job and for every individual print lane (Figure 1.12). An AVT camera then reads the barcode and verifies the printed output (Figure 1.13).

The system links to the original PDF for image comparison and for step and repeat instructions. Taken together, these systems eliminate manual set-up time for the camera.

These developments point to a future where printing presses become self-managing units, where everything from production planning, to consumables ordering, to predictive maintenance, are all generated from the press itself, or though integration with MIS systems. This will have a massive impact on label press productivity, freeing up skilled personnel to focus on developing the business, rather than managing manufacture.

Such future plants could work 24 hours a day, seven days a week, automatically tracking work process across perhaps multiple sites – and all managed remotely through a secure web environment. How long before human-free, fully-automated label production, perhaps totally controlled through WiFi and robotics, comes to the market? Probably not too long.

Workflow automation will be discussed in more detail in Chapter 9.

A TYPICAL WORKFLOW SYSTEM

This book is primarily concerned with the role and functions of label and package print industry business management information and related automation solutions, such as those offered by Label Traxx, Cerm, Tharstern, SolPrint, EFI Radius, Imprint-MIS, CRC Information Systems, OneVision, Quick Brown Fox, QuadTech, and others.

MIS systems today enable everything from materials purchasing, to production, distribution, invoicing and accounts information to be held in one computerized system, integrating with other applications, including desktop productivity tools, press management control systems, inspection, finishing, inspection, accounting and administration systems and supply chain integration.

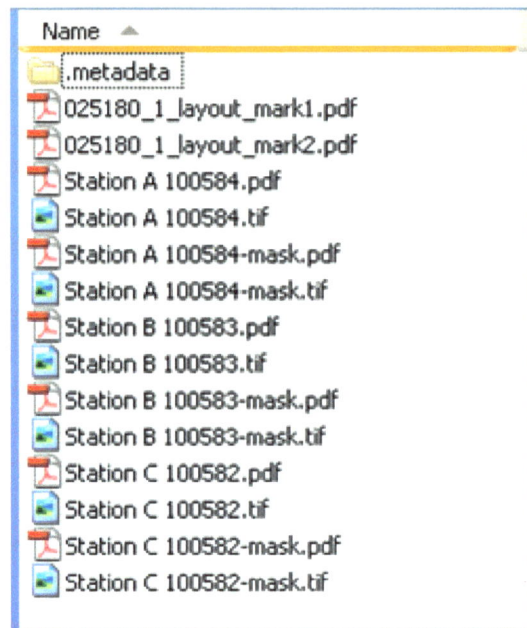

Figure 1.11 Cerm MIS creates the AVT inspection print frames

Figure 1.12 Creation of inspection files per print frame , and the printed barcode of the print frame. Source: Cerm

Figure 1.13 The printed barcode picked up by the AVT Camera

Systems are fully integrated so data only has to be entered once and changes ripple through to other modules automatically. For example, if a customer changes the quantity required on an order, the computerized system automatically adjusts the works order, re-times the job, adjusts the production schedule and stock control - and then re-calculates the quantity of raw materials allocated.

At the hub of an MIS system is a database that holds all the details required to manufacture, stock and sell the label or printed package products, including press requirements, tooling, colors/inks, plates and finishing. Materials order/stock modules include bar code-driven traceability and details of substrates. An origination module provides full control of design, proofing, platemaking, etc; the specification of the label drives the automated production planning process.

Although based on standard components, bespoke programming is also often involved in tailoring individual MIS systems to an individual company's needs. Modern MIS systems operate in real time, allowing the operator to move from window to window and enabling electronic commerce through flexible and adaptable architectures.

The following chapters aim to amplify the basic functions of MIS and Workflow Automation that have already been outlined, and examine how they relate to each other in a typical MIS workflow system (see Figure 1.14).

The overall aim is a single system able to manage the entire business, streamline the administration process and reduce costs, eliminate errors from the re-entry of data and minimize personnel. It should also be label industry or package printing industry specific as required.

A WORD OF CAUTION

As computerized data and management information continues to grow there is a danger that label and package printing companies keep hold of files that are no longer required or necessary – 'just in case' they might one day be needed. This information that companies collect, process and archive throughout their business activities, but nevertheless fail to actually use, is termed 'Dark Data'.

This can have a significant impact on company IT budgets, with companies paying for cloud storage and to meet data protection and other regulatory requirements.

A proper information management system is required to prevent the storage of too much outdated and unnecessary information. This will include the following elements:

- Firstly, develop policies and continuous training for all staff about managing data (including data on local drives, laptops, shared drives, removable devices and mobile devices).
- Understand that although all data can be stored, it should not be done without valid reasons.
- Do not keep data longer than necessary unless there is benefit to be obtained.
- Introduce a cost charge for storage. If it's free there is little incentive to remove it.
- Have a centralized e-mail management system.

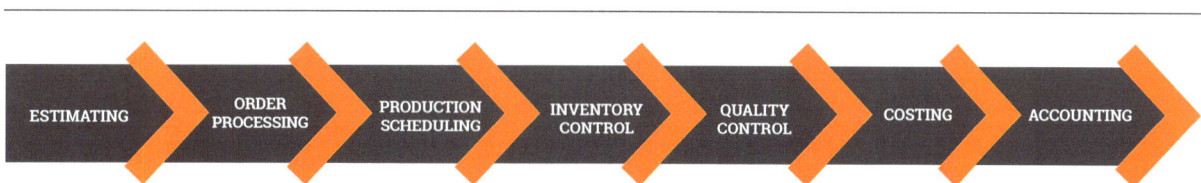

ESTIMATING → ORDER PROCESSING → PRODUCTION SCHEDULING → INVENTORY CONTROL → QUALITY CONTROL → COSTING → ACCOUNTING

Figure 1.14 A good MIS will provide a streamlined and seamless administration process of jobs through a label or package printing plant

- Delete e-mails as defined in the company's organization policy.
- Beware of data accumulated – perhaps in the cloud – by 'creative departments' without contacting or liaising with IT.
- Have a defensible deletion policy, bearing in mind legal risks associated with storing too much data.

INTEGRATION IS THE ANSWER

With end-to-end integration from estimating through scheduling, inventory, production, quality control, shipping and accounting, MIS software and solutions today facilitate the easy collection and sharing of information. A company's information collected and shared in one dynamic database will promote effective sharing and communication between departments, among employees, and even externally to and from valued customers.

The use of MIS 'smart' software in a modern label or package printing plant is able to automate workflows and optimize plant operations by recommending and planning the most efficient job routes, schedules, and allocation of resources.

Chapter 2

Estimating for label and package printing

Accurate estimating is at the heart of every successful label or package printing operation. After, all, one of the first things a prospective customer wants to know is how much is the job going to cost. Get it right and the company will make an acceptable profit on each job produced. Get it wrong and the company may regularly lose work or make a loss on the job they have produced. As a consequence, they can soon fall into a loss making situation.

Historically, the estimating process was virtually a manual operation, and the estimator had to be reasonably good at mathematics, be orderly and accurate, methodical and legible with figures – with a good technical knowledge of all the processes, the sequence of pre-press, press and finishing operations, and be able to understand what job requirements might cause production problems.

It was also necessary for the estimator to compile records of actual outputs for different operations and processes, to prepare output tables, and to regularly compare estimates with actual costings. They would have a printed estimate form which had to be filled in as the estimate progressed. Although these stages were not too difficult for a good estimator in general printing, it was a time-consuming and onerous process, with no industry standardized estimating forms for label and carton production. It was the advent of the computer and the development of dedicated estimating software that began to change things.

Estimating software for the general printing industry has helped make the production of estimates for more standard print items much easier and more accurate. Label and package printing however, can

be far more complex. There are frequently many more printing and finishing operations to be taken into account – flexo, offset, screen, letterpress, combination process, digital, hybrid, die-cutting, foiling, embossing, varnishing, slitting, web inspection, re-winding, etc. Then there is the whole area of origination, pre-press, plate and die-making, distortion for sleeving, cutting and creasing for cartons.

Even the possible range of materials that are used in the world of labels and package printing is way beyond that of the commercial printing sector. Just think of the many different paper and board materials used: matt, gloss, coated, uncoated, high wet strength, and much more. Then there are all the different filmic and non-paper substrates, such as PET, OPP, PVC, PE, OPS, metalized film and metallic foils, etc., that may be specified, as well as a whole range of different adhesive types and release liners.

Label and package printing has become even more complex in recent years with the introduction of digital printing presses, both electrophotographic and inkjet, as well as hybrid digital/conventional presses for specific types of work and applications. Frequently today, the label printer may also wish to estimate a job for both conventional and digital printing and work out the crossover point in terms of run-length and profitability.

As the starting point for any new job it is essential that the data can be carried forward to each subsequent stage – order processing, production scheduling, inventory control and purchasing , job costing and accounting – as seamlessly as possible, without re-keying or additional data entry. This can be seen in Figure 2.1.

So let's now look in more detail at the requirements of an estimating system, both in terms of what it should be expected to do, what it must include, and how it links with other modules in the MIS process.

RECOVERY OF COSTS

Bring all of the estimating requirements together and it can soon be appreciated that to produce an accurate estimate – or crossover estimates if multiple processes are being used – the label or package printing converter needs to know the cost of every material, every stage, every operation, every hourly pre-press, press and finishing operation rates before an accurate estimate can be produced. It may even mean preparing alternative estimates using different materials or processes for comparison.

Once entered, the information will need to be in a format that can also be carried through to all other subsequent modules with minimal or no additional data entry as required.

Remember too, that every label and package printing business has to pay wages, buy materials and meet expenses. These three basic items comprise the cost of a business and it is from the charges, made against individual production jobs, that costs have to be recovered. Costs that need to be recovered in the estimating and costing system through hourly cost rates, materials expenses,

outwork, etc. include:

- Cost of running offices and factory
- Capital investment – interest, depreciation
- Energy consumption
- Personnel costs – management salaries, canteen, etc.
- Wages, including holiday payments and employer's liability insurance
- Plant and factory maintenance
- Administration costs – telephone, postage, auditing
- Transport and shipping – carriage, vans, etc.

With such a complexity of estimating factors and costs, it is not surprising that the first steps in the development of dedicated label (and carton) software were being undertaken as far back as the early 1980s. Today, sophisticated Management Information Systems incorporating label and/or package printing estimating programs are available from a number of specialized vendors, such as Cerm, Label Traxx, Tharstern, Optimus, EFI Radius, CRC and Globe-Tek. Pretty well all the MIS systems today are also integrated with one or more partners, including the latest Esko offerings.

Even with the sophistication of the latest MIS systems, estimating remains at their heart. Being able to estimate for different conditions largely dictates whether the MIS can readily support multiple print processes and formats, such as wind direction or the nesting or staggering of label or cartons. Many software suppliers get around this by integrating with CAD systems that can provides such calculations – in some cases being able to create nesting within the estimating program.

As the number of jobs increases and average run lengths get shorter and shorter, so the printer must be able to quickly create profitable estimates. So how

Figure 2.1 Estimating is the starting point in a streamlined and seamless MIS workflow

ESTIMATING ORDER PROCESSING PRODUCTION SCHEDULING INVENTORY CONTROL QUALITY CONTROL COSTING ACCOUNTING

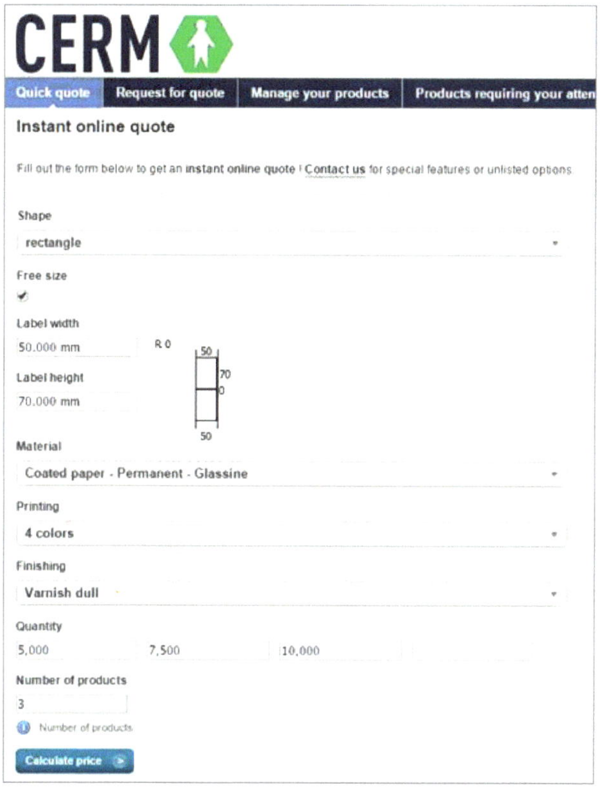

Figure 2.2 Instant Online quote form using Cerm MIS software

does this all come together for the estimator using today's MIS and other estimating software?

REQUESTING A QUOTE

The request for a quote – in a standardized format – should describe the external characteristics of the label or pack (e.g. size/shape, wind direction, number of colors, paper, quantities, etc.) and may need to be for different quantities, perhaps with different print or finishing options, while the converter may also be looking at printing the proposed job by either conventional or digital processes – whichever offers the best return. An example of an Online quote form using Cerm MIS software can be seen in Figure 2.2.

As can be seen this requires data to be entered for shape, size of label, material to be used, number of colors and any finishing, quantity and the number of products, thereby enabling a price to calculated.

Some printers will already have a website that collects the necessary job information, without calculation. This is regarded as the first step.

The information is then stored and presented to the 'estimator'. The actual 'calculation' becomes step two. The calculation establishes the cost. On top of this cost, somebody in the company (as step three) will define the proposed sales price, normally by adding a margin or in evaluating e.g. the added value of the estimate. Step four is to communicate the sales price in a computer generated quote letter to the customer.

To prepare the estimate the converter may additionally need to search on screen for appropriate substrates, inks, tooling or cylinders and determine their availability. Once an estimate(s) appears on screen – with the selected wind direction and maybe different quantity prices if required (see Figure 2.3) – it may need human tuning in relation to competitors or profitability and, finally, printed onto the label or package printing company's letterheads, or sent to the customer by fax or e-mail.

If the customer subsequently telephones to discuss the quote, then it probably needs to be instantly called-up on screen for review and maybe amendment, perhaps to compare costs and pricing between different methods of production.

With some of the latest MIS software, label companies thinking of investing in, say, a new digital press or already offering conventional and digital technologies, can run dual or trial estimates for the different technologies and do cost crossover comparisons with the other presses or press technology that they are already running or are thinking of considering. This can be based on the choice of press or on capabilities for colors and in-line finishing options, or cost (as seen in Figure 2.4.).

Figure 2.5 also shows a crossover graph, this time prepared to compare estimates for printing a job either by digital or flexo processes and based on material cost, hours and different quantities. In this cast the crossover is given as 24,140 labels at a cost of £1,310.

Leading MIS software suppliers today all

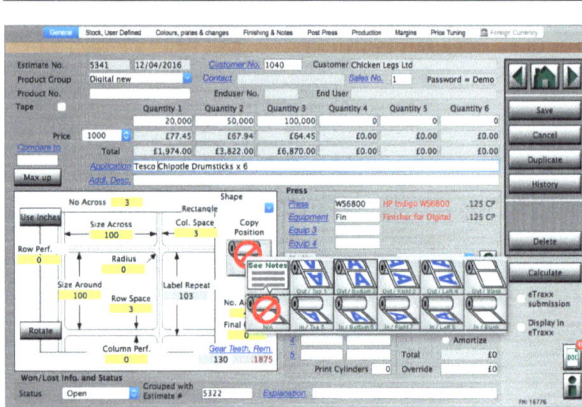

Figure 2.3 Label Traxx Online estimate process showing items such as quantity, wind direction, size, press and price per thousand

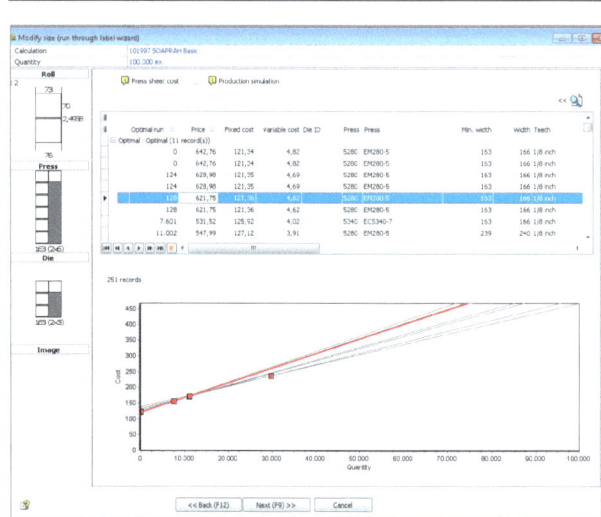

Figure 2.4 Choice of presses, based upon capabilities for colors and in-line finishing and cost. Source: Cerm

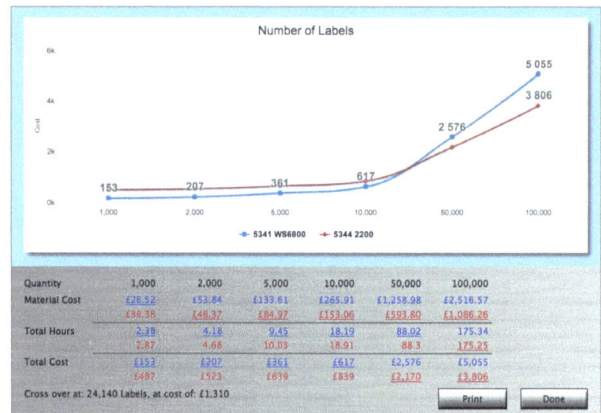

Figure 2.5 Shows a crossover graph used to compare estimates for a digital press and a flexo press. The graph illustrates the crossover point. Source: Label Traxx

incorporate an 'online quote' facility that will calculate the selling price online, or for a variant of an existing estimate (such as another quantity or process), will follow the rules already defined in an existing estimate. Intuitive estimating algorithms are able to provide multiple price options including best margins and different production routes (Figure 2.6). Existing rules may even apply for a brand new quote, but only in simple cases.

Another current trend is for the generation of 'tenders' for existing or potential customers asking for a complete pricelist for a wide range of labels in order to confirm a label printer for all of their products for the next one or two years, based upon price. This trend led Cerm for example, to write a dedicated label-wizard to make dozens of estimates in one operation.

It must be remembered however, that any estimating system is only as accurate as its cost rates. They must be up-to-date and be an accurate reflection of the true costs of running the business in terms of manpower costs, materials costs, machine running costs, ancillary costs and shipping costs. If the cost rates are not accurate, then the estimate itself will be wrong – possibly losing business or taking on work that will not make the anticipated sales margins.

REQUIREMENTS OF ESTIMATING SOFTWARE
Key requirements of a suitable MIS will include:
- Easy to learn and use, with flexibility during implementation
- Provide fast, accurate, and consistent estimates regardless of user experience

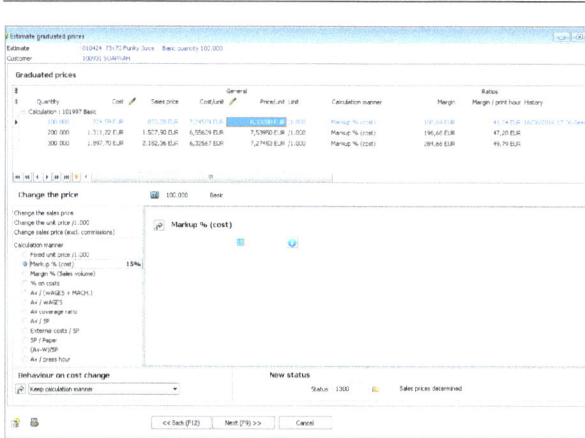

Figure 2.6 Determination of sales price, based upon cost per quantity and margin. Source: Cerm

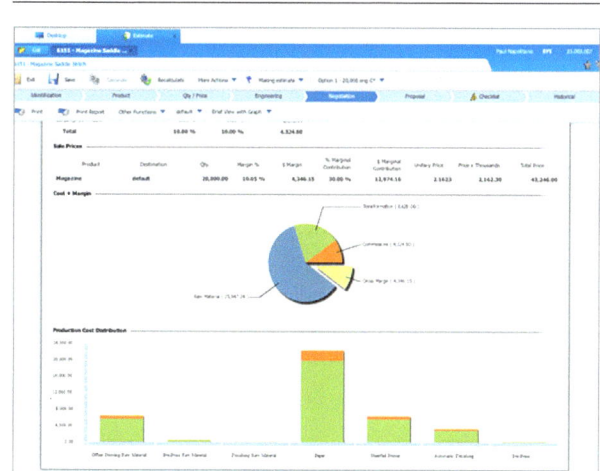

Figure 2.7 A negotiation screen in EFI iQuote Negotiation provides a visual detailed breakdown of different costs and margins within the job being estimated

- The setting of pricing restrictions on the software user. For example, restricting the use of discounts to, say, 15% or 20% for the sales manager and 10% for others
- Adding company spoilage or waste figure allowances based on specific processes or job complexity
- The viewing of substrate, ink, foil, standard cutter or other in-house stock holding before making production decisions
- The defining of sales price bands for specific types of customers or markets, e.g. trade customers, new accounts, long-term customers, etc.
- Include multiple quantity pricing capabilities, perhaps with a sales price override
- Calculate profitability targets for all or new types of work
- Provide consolidated estimating knowledge in a centralized shared database
- The ability to undertake routine estimating/quote analysis by type of work, customer, print process, machine, personnel, profitability, etc.
- The ability to generate customer estimates for conventional analogue or digital printing processes, and prepare crossover costs between two or more presses

- The updating of any historical estimate based on the latest rates and machinery
- The ability to calculate ink, varnish or coating usage based on the substrate used, expected coverage, etc.
- Provide for the tuning of pricing whilst monitoring profitability
- Ensuring that shipping/despatch costs can be included in the estimate, based on freight method, location, weight of the shipment
- Access customer estimate history

With today's sophisticated Online quote software, virtually any member of staff can produce an accurate quote, so eliminating delays in getting quotes back to customers. Indeed some software will even enable customers to quote jobs themselves.

NEGOTIATING WITH CLIENTS

With web-based estimating systems it allows sales to provide immediate answers to questions during negotiations with clients (see Figure 2.7), even to the point of providing detailed quotations on the road and upload artwork for the job.

Once estimates have been approved and

production initiated, the estimating module will also need a range of analysis and reporting options that can provide both on-screen and printed estimate analysis. The software will provide comprehensive sales analysis by customer, type of work or market sector, profitability, customer documentation, and follow estimates through to final ledger posting.

Product ID	Description 1	Production comment	Ordered quan	Gallus	HP6000	Nilpeter	Xeikon 3500
Reference calculation (with size) : Irregular (1.969000 x 6.299000) Gallus 110004 (5 record(s))			278,...	5,751.85 USD			4,402.00 USD
100016	1.969 x 6.299 Soap Ventoux - relaxation	Xeikon 3500 - 5	43,000	5,751.85 USD			4,402.00 USD
100017	1.969 x 6.299 Soap Ventoux - provence	Xeikon 3500 - 5	40,000	5,751.85 USD			4,402.00 USD
100019	1.969 x 6.299 Soap Ventoux - tonic	Xeikon 3500 - 5	45,000	5,751.85 USD			4,402.00 USD
100020	1.969 x 6.299 Soap Ventoux - Muscat	Xeikon 3500 - 5	50,000	5,751.85 USD			4,402.00 USD
100120	1.969 x 6.299 Soap Ventoux - Strawberry	Xeikon 3500 - 5	100,000	5,751.85 USD			4,402.00 USD
Reference calculation (with size) : Irregular (2.362200 x 3.937010) Nilp. 110016 (4 record(s))			87,000		1,528.81 USD	1,742.92 USD	
100066	2.362 x 3.937 PG - PHYT'S - h eucalyptus	HPI 6000 - 5	25,000		1,528.81 USD	1,742.92 USD	
100067	2.362 x 3.937 PG - PHYT'S - naturoderm	HPI 6000 - 5	22,000		1,528.81 USD	1,742.92 USD	
100068	2.362 x 3.937 PG - PHYT'S - hc circulation	HPI 6000 - 5	21,000		1,528.81 USD	1,742.92 USD	
100069	2.362 x 3.937 PG - PHYT'S - hc drainage	HPI 6000 - 5	19,000		1,528.81 USD	1,742.92 USD	
Reference calculation (with size) : Irregular (3.149610 x 3.346460) Gallus 110030 (3 record(s))			21,000	788.59 USD	506.46 USD		
100099	3.150 x 3.346 Seigneur de Lauris - GIGONDAS 2006	HPI 6000 - 3	7,000	788.59 USD	506.46 USD		
100100	3.150 x 3.346 Seigneur de Lauris - VENTOUX 2007	HPI 6000 - 3	7,000	788.59 USD	506.46 USD		
100098	3.150 x 3.346 Seigneur de Lauris - VACQUEYRAS 2009	HPI 6000 - 3	7,000	788.59 USD	506.46 USD		

Figure 2.8 The ability to choose the cheapest alternative for every group of 'similar' products (green is the cheapest option in this screen shot). Source: Cerm

ADVICE AND CAUTION

While the estimating and quote process has become ever more sophisticated and automated, the estimator's role in a label or package printing plant has not been superseded. A good estimator may be able to suggest alternative options (see Figure 2.8.) while working on an estimate.

The rule should be to quote as exactly as possible for what has been asked for, but also be able to – under an appropriate heading such as 'additional work suggested' – any alternatives or improvements which it is thought might interest the customer. Each suggested item can be given a reference so that the customer can come back to decline, discuss or accept. The estimate should be carefully worded to indicate clearly to even a non-technical customer, exactly what has been included.

Together with the estimate, either on the back of the quote or clearly referred, there should be set out the company's Standard Terms and Conditions.

If a printed job is to be shipped on a periodic, monthly, quarterly or long term basis then the estimate should include terms for 'Payment on Account', or a sum added in the estimate to cover interest.

The best label and package printing estimating software will literally process thousands of pieces of information for each job, including a careful look at virtually every piece of equipment and capability found in-house. It will capture the best practices, giving any estimator the ability to perform at the highest level, regardless of experience. The software protects the equity found in the most experienced employees, while enforcing the business rules already developed by the company, making estimates accurate, consistent, and profitable.

When label and package printing businesses are confident in their estimating abilities, producing results that are consistent with their estimating process, then they are most likely to generate higher margins, and have the knowledge at their fingertips to understand how competitive they can be without losing money on a job.

For the future, the trend is towards ever greater estimating software integration – with other industry partners and with cloud computing – as well as the way that software is brought together and used. Rather than just buying software, the future may be more to do with licensing software and estimating modules that include after support and other services.

Chapter 3

Order processing and job management

Once a label or package printing estimate has been agreed with the customer it will then need to be efficiently and accurately turned into a product or job order, with the job passing from sales into production and eventually on to despatch to the customer. The aim of an order processing or job management module or software is to automatically convert the estimate or sales order information into the production order – that is to create the job bag or work ticket – with all the information being viewed on-screen (or printed out), and accessible to everyone in the plant, without the need for any re-keying of text or data.

The MIS order process or job management module will then be able to track the entire life-cycle of the job, from order entry through order processing and job management – including integration today with pre-press, color management, printing, inspection and finishing – to accounting. Each job will travel through the production workflow, visible to anyone in the organization, with managers or operators able to view the status of the job, and where it is in the production cycle, without leaving their desk or work station.

Depending on the system, orders may be raised from estimates, from previous orders (e.g. reprint work), from customer-facing web portals (such as FRONTDESK), through Electronic Data Exchange (EDI) or manually. If an order is raised from a previous order or an estimate, then again no re-keying of data should be necessary. Re-orders can be quickly expedited with a simple job duplication function. Job creation is automated

based on previously specified technical and commercial information. Auto-planning features can enable similar labels, or labels with different shapes on one plate, to be ganged together. This data can then be sent to pre-press systems such as Esko's Automation Engine, for automated pre-press, step-and-repeat, ganging, etc. But more of this later.

A combined customer-facing web portal and automation system enables customers to place, track and manage all of their orders, and to transfer those orders directly into the converter's MIS and production system automatically. Custom product categories can be set up for each customer's orders, with robust file upload and optional online document editing and VDP capabilities.

With E-Commerce a company's market reach can be expanded, while customers can be empowered to fulfil ever more of their needs online. E-Commerce offers round the clock customer service tools for processing online order requests, reviewing job status, tracking shipped jobs, and reprinting invoices. By providing client relationship services and tools online, skilled personnel are free to concentrate on revenue generating activities.

If Electronic Dara Interchange (EDI) is integrated with a company's MIS, then computer-to-computer transfer of standard business documents in a standard electronic format from one company to another can be achieved without manual intervention.

Several documents are produced directly from the order, namely:

- The **job bag**, which lists works instructions, printing information and any outwork requirements.
- The **work sheet or work ticket,** which lists every item of labor, materials to be used and outwork on the job.
- The **order acknowledgement**, which prints, faxes or emails a letter that can be sent to the client accepting their order.

Orders may have additional areas to enter work-specific instructions, delivery information and client reference numbers and it may also be possible at this stage to set pre-press and proofing schedules, as well as target dates against the job. The ideal situation is to use a very flexible software tool for entering, tracking, and interacting with production data in real time and available to everyone in the organization using a simple web browser (or tablet) interface. This is becoming ever more important.

Market pressures in the label and package printing sectors today, mean that the order life cycle is growing shorter and shorter. Printers and converters are being pressurized to produce more and more jobs, in smaller quantities, with the same

people, presses and ancillary equipment. This requires working systems that orchestrate and link together the different production and workflow stages that include estimating, scheduling, quality control, shipping, online web ordering, customer proofing and approval, file management, and prepress production.

Before proceeding further, it should be noted that there are important differences between MIS systems developed for the commercial print market and those developed specifically for labels and packaging converters. These include:

- In label and packaging printing, estimates are often made for a longer period of time, e.g. one or two years, and frequently contain the pricing rules (staggered prices per 1.000) and the technical alternatives (conventional/digital)
- The creation of products in line with the estimate may take place before the first order for a product comes in. The aim is to get prepress done before orders are there. This is all about product creation and approval
- When sales orders come in, these are nothing more than product, quantity, price, shipping details. In other words, an 'entry' point to decide what to do, which can be one of three options:

1. Purchase the products (e.g. sales orders for blank labels that are not directly produce in-house).
2. Take the products out of stock (e.g. when they are produced in larger quantities and shipped in smaller quantities, but retained in stock in between).
3. Produce the products
   - For stock
   - For immediate delivery.

An order confirmation will be sent.

Figure 3.1 The place of Order Processing within the MIS workflow process

Only in case '3' above will there be production jobs that require time and material.

Many MIS systems can only work with 'jobs': they can define the different products within a job but they are stuck with partial production/take out of stock and with the purchase of products.

So the function of creating a label or package printing job is to:

- Group all sales orders with products 'to be produced', based upon the same estimate
- Find the cheapest production route among the production alternatives within the estimate
- Combine (gang) labels on frames
- Create new job-ID's

This means that it is sometimes not one estimate that is 'turned into' a job. The estimate will be re-used several times for job creation and will, in this role, be the source for production, to be improved after every production cycle when there is quality control feedback, therefore continuously optimizing for repeat production.

Only after all the order processing and job information has been entered into the system does it becomes possible to start production planning and scheduling, assign job priority, track jobs in progress, view press or process capacity, track and maintain materials inventory using barcodes, monitor quality control, as well as order and receive materials electronically – all on-screen and without leaving the desk – and finally taking the job through to costing and into accounts. Order processing is therefore a key stage in the workflow process (see Figure 3.1).

Essentially, an effective MIS order processing module becomes a centralized hub that captures, holds and displays all the information needed about each particular job. Any person in the management or production team should be able to view any job component or production stage and be able to instantly view any job in the process of production at any time.

Order processing software can go on to perform a number of additional functions: automatically raise a cost sheet and record times against a job; prepare and issue delivery/shipping note documents; and print out the required pallet, carton or shipping labels

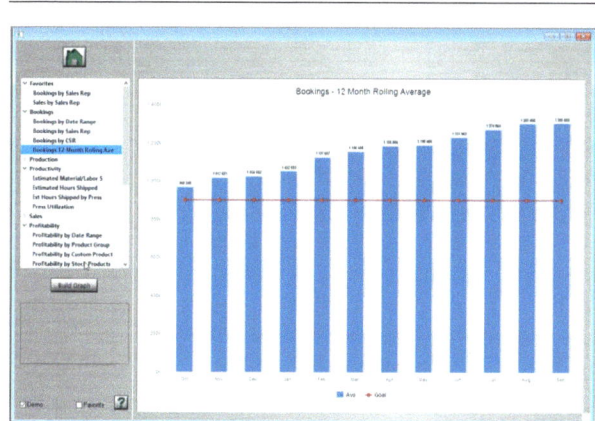

Figure 3.2 A 12-month rolling average of bookings. Source: Label Traxx

– whether for a single job or delivery of multiple jobs to different clients. The system should also be able to maintain a history of orders in a quick and easy to navigate list which provides a full searching and filtering facility.

Built-in production reports can help to keep track of all active orders, bookings (see Figure 3.2), pending proofs and invoices, required materials, press job allocations, delivery response times, sales turnovers and much more. If required, an extensive drill-down view of all aspects of any particular job, or group of jobs, can be created.

High-speed database search facilities, with some MIS software offering color highlighting to identify features, machines, substrates or personnel can be extremely effective. Sales and order histories in a system can be broken down by client, type of work or sales representative so as to calculate sales commission figures or perform detailed analysis of orders.

It is also important that the MIS job management module/software enables the tracking of all production extras such as 'author corrections' with an automatic reminder for inclusion during charging. It may additionally have an 'extras authorization' facility for additional work identified prior to commencement or unexpected cost that have arisen – such as artwork amendments, additional proofing

– as well be able to compare the estimated and actual job costs so that variances can be highlighted or reported. It is also possible to generate reports that show orders/ bookings over a period of time as required. See Figure 3.2.

When booking the job in from an estimate, a good order processing system automatically looks into the stock modules and allocates all the necessary materials to the job. If the materials are not in stock or unavailable, the software will raise a purchase order to the preferred supplier. Stock allocations and purchase orders can be modified at any stage. However, the system should not let an order be forgotten or forget to allocate materials to a job – a valuable safeguard when under pressure.

FEATURES IN AN ORDER PROCESSING/JOB MANAGEMENT SYSTEM

Put together, some of the key features that a printer/converter can expect to find in a good order processing or job management system include:

- Creating and printing order acknowledgements
- Assigning job priority
- The tailoring of work instructions against requirements
- The ability to provide electronic on-screen or hard copy job dockets
- Being able to automatically raise purchase orders for any required materials or outwork
- The automatic display of customer notes

- Drawing attention to additional charges
- Materials information (availability, pricing, etc.)
- Providing job docket management
- Timesheet posting and operator performance analysis
- Enabling job quantity amendments and automatic re-calculation of estimate or production data
- The facility to import and highlight order changes
- Customer and supplier information
- Manage the product approval cycle
- Initiate production planning (drive pre-press)
- The ability to provide work in process analysis and reporting
- Integrating with finished goods stock
- Automatic job creation from an XML or CSV file
- Preparation of unlimited production notes against estimates or jobs
- Providing delivery schedules and consolidated delivery notes
- Analysis of sales and order histories and the creation of reports, including WIP, added value, profitability, overhead and budget recovery
- Preparation and issue of delivery notes
- Printing out shipping labels
- Create detailed work-in-process analysis

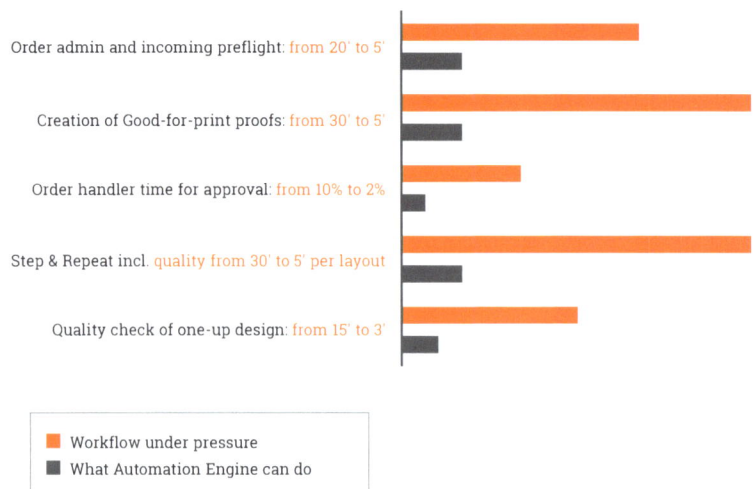

Order admin and incoming preflight: from 20' to 5'

Creation of Good-for-print proofs: from 30' to 5'

Order handler time for approval: from 10% to 2%

Step & Repeat incl. quality from 30' to 5' per layout

Quality check of one-up design: from 15' to 3'

■ Workflow under pressure
■ What Automation Engine can do

Figure 3.3 Shows what can be achieved with automated pre-press in Esko's Automation Engine

INTEGRATION BETWEEN MIS AND PRE-PRESS

Today, MIS is increasingly being integrated with pre-press jobs and production jobs. After all, it seems logical that pre-press will need to be involved with any new product emanating from a sales order. Pre-press is therefore automatically informed and a JDF sent to, say, Esko's Automation Engine, so as to create a 'pre-press job' with the same identity. Where a product is a reprint – without any changes – this step can be skipped. Note: Automation Engine Connect is a toolkit that allows integration from a third party product with an Esko component (WebCenter, Automation Engine or ArtiosCAD).

A job in Automation Engine represents a production order that organizes the data storage for the job but also the job's metadata, its link to order ID, due date, customer info, Customer Service Representative contact, etc.

Besides these administration attributes, a job can also contain the graphical specifications like barcode, inks, RIP options and so on. This job information can be used in any workflow to take advantage of all the data that is already there, avoiding double entries.

With pre-press jobs, the MIS asks the pre-press department's Automation Engine to create a new product based on the technical information provided in the estimate. New artwork for the job may then be delivered, with pre-press making it print ready, proofing it and obtaining approval for printing. When a new product is created in MIS, it automatically creates a 'pre-press job' in Esko's workflow server Automation Engine. Both the MIS and Automation Engine always look at the same pre-press information. Time savings can be quite significant, as can be seen in Figure 3.3.

Some solutions may also offer 'production jobs' in prepress the facility to undertake step-and-repeat, RIP-ing and plate making, the sending of jobs direct to a digital press, or to gang or combine (Figure 3.4) similar labels or labels with different shapes on one plate layout, with data being sent to Esko's Automation Engine for automated step and repeat.

When proofs are ready to be sent to customers in the integrated MIS job management/Automation

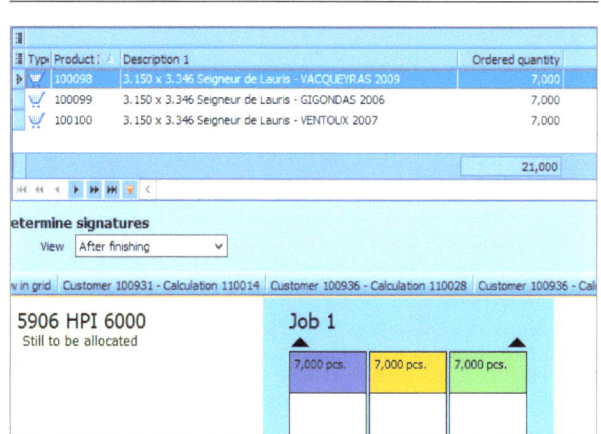

Figure 3.4 Combining labels in lanes, next to each other to save production switch times (digital or offset). Source: Cerm

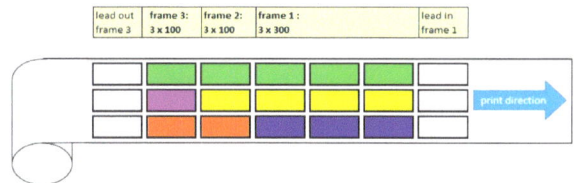

Figure 3.5 The use of identification print frames in between the good labels

Engine the MIS selects all the 'Proof ready' products and e-mails them to the customer for approval, automatically attaching the proof to the outgoing e-mails.

The JDF created by the MIS can additionally contain Lead-in Lead-out (LILO) information, enabling a job information barcode label or identification 'print frames' to be printed at the start and end of the job (Figure 3.5), or print further LILO labels containing the cutter color so that print clicks can be saved for off-line finishing.

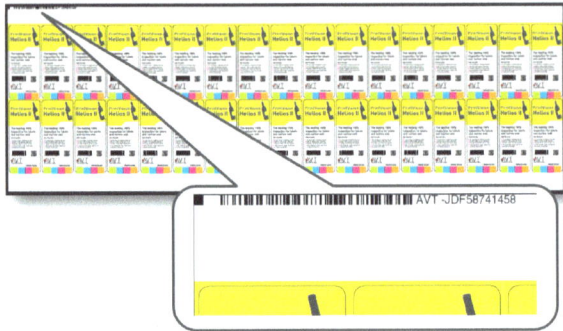

Figure 3.6 AVT Zeroset system uses a printed barcode for verifying the output

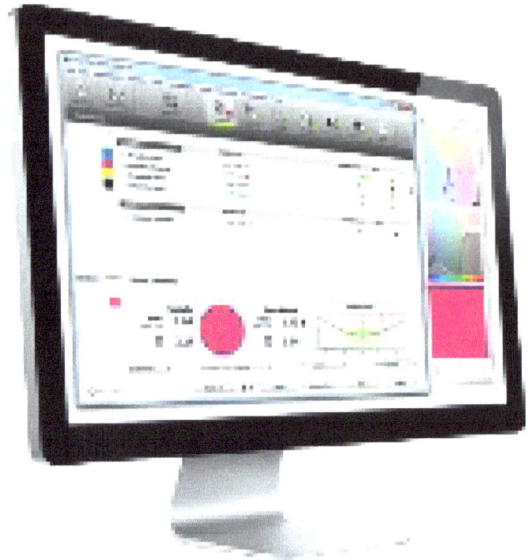

Figure 3.7 X-Rite's ColorCertProduct Icon Desktop Tool

MIS DRIVING PRINT AND FINISHING

The impressive rise of digital print for labels has pushed major developments in both MIS and pre-press software. Even to the extent that Step-and-Repeat functionalities are now available in many Digital Front Ends: Gallus LabelFire, HP Indigo, Xeikon X800, Screen, etc. However, it's not just digital printing that benefits from this integration. Flexo presses can also benefit from an MIS interface. The Cerm- Gallus press-link for example, can drastically reduce the set-up time of the press and will feed back information to the MIS.

If press set-up time can be drastically reduced, what about inspection and finishing? It perhaps seems obvious that when the small-run problems in digital print disappear, then a new bottleneck in finishing soon becomes apparent.

One of the first to benefit from an electronic MIS/production interface was AVT Camera Inspection. With a link to the original pre-press PDF for image-comparison and instructions for Step-and-Repeat, AVT can now reduce the set-up time of their camera inspection systems to zero.

The MIS creates the print frames, while pre-press creates inspection files per frame (in an AVT folder). The press prints the barcode of the print frame, with the AVT camera reading the barcode and verifying the output. See Figure 3.6.

AVT have also introduced their 'iCenter Platform',

a cloud-based technology that provides tools to set cross-site quality standards, auto-analyze PDF files for inspection, and extract business intelligence from the production floor with a seamless connectivity to MIS and pre-press solutions for optimized automated workflows.

Another early innovator of MIS integrated production technology was ABG International. The manual set-up time of their slitter-rewinder for labels has been reduced to less than 1 minute with an MIS interface in which the operator retrieves MIS job instructions on the slitter, which provides all technical information, and for an automated set-up of the slitter knives.

In other developments, X-Rite Incorporated, a global leader in color science and technology, and its subsidiary Pantone LLC, have recently introduced the latest version of ColorCert Suite 2.7. This Suite of software products is now able to provide real-time color and print quality process control and reporting (see Figure 3.7) to help packaging printers and converters better manage the complexities of CMYK, extended gamut, and spot color workflows,

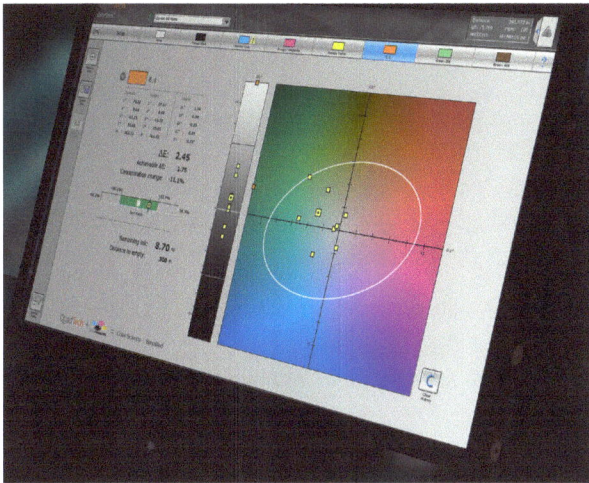

Figure 3.8 QuadTech's ColorTrack automated color management solution

Figure 3.9 Monitoring shop floor data collection. Source: Label Traxx

regardless of the printing process, substrate, or industry standard. It also includes a number of connectivity features that make the communication of color specifications and performance metrics easier and more streamlined than ever; from pre-press and the ink room to production process control, it helps organizations manage precise color and print specifications, uniting traditional silos of information into one common workflow.

Also introducing new color management software is QuadTech. Described as a unique, highly adaptive new solution for packaging applications, ColorTrack operates without any hardware modifications to the press, integrating with ink formulation software to simplify workflow and reduce the number of ink corrections needed to achieve accurate, optimal color.

Described as a 'color expert in a box' ColorTrack (Figure 3.8) automates the process of delivering absolute consistency from press-to-press, shift-to-shift, and plant-to-plant', and is said to be the only software that offers such a level of press-side connection between color management and ink management.

SYSTEMATIC JOB CONTROL

The overall aim with an integrated MIS order or job processing system is to know and understand where every label or package printing job is in the workflow process and to provide for systematic job control.

The system should offer quick and easy database searching and flexibility of job viewing and provide a clear chain of custody process that is accessible to everyone in the factory and which tracks and monitors every stage of production across the shop floor in real-time and throughout the entire lifecycle of a label or package printing job. See Figure 3.9.

Some of the job management software solutions on the market are able to configure their individual workflow requirements to address the specific production management needs for a range of products that they offer, including labels, shrink sleeves, in-mold labels, flexible packaging, blown film extrusion products and folding cartons.

These technologies enable the entire production workflow to be integrated, standardized and substantially automated. They also cover detailed quality control and optimization of print files and images using RIP and associated software, as well as individual solutions for the imposition of PDF files

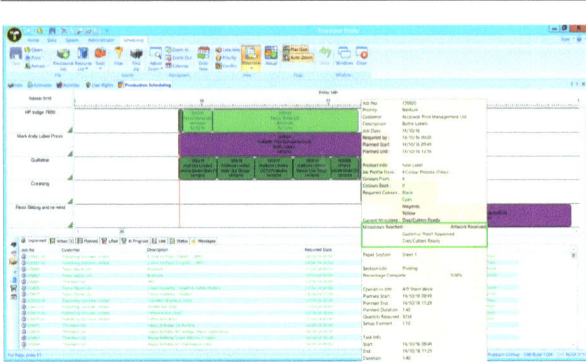

Figure 3.10 Press operators can view milestones for scheduled jobs direct on the shop floor. Source: Tharstern

or flattening of transparencies – all done automatically.

Print data is automatically subjected to a quality control and then optimized. The automated production of die-cutting moulds or laser-cutting configurations, standard cut lines, a white background and the automatic dispatch of a release PDF to the customer for approval saves labor time and staff resources.

If time sheet posting is provided then it becomes possible to also capture operator activities, including wage rate and appropriate shift, to allow accurate costing of jobs and performance analysis of operators. There may also be options for the estimated labor, material or outsource costs to be automatically apportioned to a job following production, enabling complete costing analysis.

Depending on the MIS being used, press operators may have the capability of consulting their works or job schedule per machine, view milestones for scheduled jobs (Figure 3.10), and even enable them to indicate what they are doing. A change in schedule will be picked up automatically on the shop floor, with no need to re-distribute work or job lists. Semi-finished goods can also be tracked.

A useful feature of job management software/ modules is the ability to assess real-time activity in a dashboard style or executive snapshot view, showing current work-in-progress (WIP), current scheduled capacity, sales, added value and overhead and budget recovery.

With some job management systems it may also be possible to have a credit checking link (to, say, Accounts Link) to ensure that customers do not exceed approved credit limits or to make sure that any jobs on 'stop' or 'hold' are identified and flagged up.

THE USE OF EDI ORDER PROCESSING

For those label converting or packaging printing companies that regularly deal with some of the large supermarket and retail groups – Walmart, Tesco, Sainsbury, Safeway – or consumer products groups, there may be the opportunity to receive orders and send invoices electronically. It may be something that they are already doing or perhaps been asked if they can implement in the future. There are certainly a number of key benefits that can be obtained by embracing EDI. So what is EDI order processing?

EDI is the abbreviated name for Electronic Data Interchange, a means of computer-to-computer transfer of standard business documents in a standard electronic format from one company to another. Quite simply, EDI orders are automatically created in the system so that no manual order entry is needed.

By moving from a paper-based exchange of business document to one that is electronic, businesses enjoy major benefits such as reduced cost, increased processing speed, reduced errors and improved relationships with business partners.

Such a system will generally need to be set-up with a customer's head office IT department but once established and configured to the needs of both participants, then job orders, invoices, etc., can be generated and transferred automatically. Confirmations and alerts can be sent through the email system to keep everyone informed of what is happening or not happening.

Accuracy is improved with EDI. Time is saved as automated systems can send data at any time of the day or night so that orders will already be in the system whenever the management or production

Figure 3.11 CRM software provides a means for businesses to manage sales, marketing, orders and support services with customers

personnel arrive.

EDI replaces postal mail, fax and e-mail. While e-mail is also an electronic approach, the documents exchanged via e-mail must still be handled by people rather than computers. Having people involved slows down the processing of the documents and also introduces errors. Instead, EDI documents can flow straight through to the appropriate application on the receiver's computer (e.g., the Order Management System) and processing can begin immediately.

CUSTOMER RELATIONSHIP MANAGEMENT (CRM)

Customer Relationship Management (CRM) solutions are not new and, when used correctly – or even integrated with an existing or new MIS – can significantly enhance a label or package printing business. However, there are many different systems and software packages on the market, including bespoke software, and it can be difficult to choose

the right software for a business. It also needs to be a system that grows with the business.

The overall aim of CRM software is to provide a means for businesses to manage their customer orders and data and their interactions with customers, to automate the sales, marketing and customer support operations (see Figure 3.11), to better manager employee, supplier and business partner relationships, as well as providing access to business information. It can also be used to manage business contacts, sales leads and successful orders.

Used well, CRM software will enable businesses to provide better customer services, aid sales teams in cross selling, enables sales personnel to be more effective, drive enquiries to a company or sales website, assist in closing sales orders, offer a means of retaining existing customers. Importantly, CRM systems will enable marketing, sales, customer service teams and management to have a better understanding of their customers and what they expect. Automated reports can be prepared and available to both users and non-users.

Additionally, if they make use of a mobile CRM app (such as that available from Tharstern) they can be fully integrated with the MIS to provide all the information that they need about each and every customer whether they are in the office, out on the road, or visiting a customer facility. Dashboards are available to view all data in one key central location.

Label and package printing businesses can install CRM software in their own internal network facility or, today, can have access to web- or cloud-based applications, accessible from almost anywhere, on any device, in which the required software is hosted by an external provider as required. Integration with existing MIS and data systems is usually quite simple.

Chapter 4

──────

Efficient job planning and production scheduling

──────

The process of planning jobs and managing their production through a label or package printing plant can be quite complex. The process need to constantly monitor the changing workload of individual machines and printing processes, complete print production orders to schedule, and maintain accurate staffing levels. Ideally it should also highlight spare capacity, display slippage, spot bottlenecks or problem areas, offer 'what if' scenarios and prioritize orders thereby maximizing the company's technology investment and minimizing machine downtime.

──────

The growing proliferation of shorter runs, the need for multiple versions or variations, a requirement for sequential coding or numbering, jobs with hot or cold foiling, embossing, matt or gloss varnishing, are just a few of the things that can challenge scheduling. Multi-shift and multi-process pressrooms present further challenges. Even pressrooms with just one or two presses and a handful of jobs at any given time, can make effective scheduling tricky.

It is not simply a case of lining up jobs by due date, in the order they were received. Label and package printing has a great many variables to be considered in order to strike a good balance between efficiency on the pressroom floor and shipping the printed jobs when promised. Indeed, the most efficient job sequence is not always readily apparent.

Multiple deadlines, urgent rush jobs, special finishing requirements, delays in ink or paper delivery may all make it necessary to arrange jobs in scheduling sequences that are far from ideal with respect to production efficiency. Batching jobs by material, size, or ink can mean that jobs with due dates that are even days or weeks later may be scheduled to print before those with closer due dates.

Indeed, there are often so many variables, especially those that may occur unexpectedly – employee sickness, press breakdown, damaged plate or cutter, temporary shortages of materials, conflicts with jobs requiring the same plate cylinders, the need to delay one job so that another may be printed on a particular press – that any number of criteria may have to be constantly evaluated in order to determine the order that labels or packaging should be printed in. Ideally, the scheduling system should also be able to block-out recurring or planned maintenance periods where a press or process is unavailable.

All these requirements and pressures lead to a need for an efficient and easy way of conveniently presenting and viewing jobs that need to be printed in some form of job planning or scheduling system. A system that provides an easy-to-view means of sorting, grouping, controlling, planning and viewing the sequences and steps of production on possibly an hourly, daily, weekly, shift, machine or personnel basis as required. It can also be valuable for the scheduling system to undertake capacity analysis over a defined period to determine potential future bottlenecks.

Modules used for job planning and production scheduling in the label or package printing plant are today an essential element of a MIS workflow system and can provide full visibility of machine capacity, both short and long term, and offer detailed scheduling of each production step of the job, from customer approval right through to shipping. Depending on the software it may also be possible for press operators to consult their work schedule on their press or feedback what they are actually doing. A change in schedule can be automatically seen on the shop floor.

Increasingly, there are also MIS systems that are now starting to offer real-time feedback of shop floor activity to the scheduling board or screen via job tracking terminals or JDF capable systems. Others may look to define the current running efficiency of a resource for new operators, or identify machine faults to automatically adjust running time of operations. Some can provide an out of sequence alert, indicating jobs started early or not as planned, and so alerting the scheduler.

So let's look at some of these production scheduling systems in more detail.

PRODUCTION SCHEDULING SYSTEMS

Production scheduling systems or tools come in a wide variety of formats and systems starting from spreadsheets generated on a PC screen or printed out onto paper, and ranging through White Boards, Magnetic Boards, Card Planning Boards, Adhesive Strip Boards, right up to sophisticated production scheduling software incorporated into the latest MIS systems.

All of the above systems, sometimes used in combination, can be found in the industry. Why? Because, while computerized production scheduling systems have definitely become an essential planning tool, they frequently only provide a limited view of the

production schedule, failing to provide a planned 'Big Picture' or total production schedule overview that everyone can see. Constant scrolling or passing round instantly outdated printouts can be of limited help to busy production departments who must have a complete overview of the rapidly changing production schedule to make any informed amendments.

A guide to some of the main types of production scheduling boards and computerized systems that are available to the label and package printer are therefore briefly outlined below:

- **Magnetic Planning Boards**
  Magnetic production planning board kits provide a wider format picture enabling a complete production schedule to be viewed at a glance. They can be rapidly updated, give an overview of production and presses, and may be operated as stand-alone scheduling boards or alongside software systems.

- **Card Planning Boards**
  Card Planning Boards are an ideal system for when a large amount of job information is required to be stored and shown. A range of different card capacity sizes and heights are available. Card boards are ideal for production, project and personnel planning and are available in one or two week systems.

- **T-Card Production Planning Boards**
  Job information is written or printed onto a T-shaped job card (T-card) body with the 'at a glance' information onto the visible card shoulder. Information can be handwritten or transferred directly from a computer printer or photocopier using perforated job card sheets.

- **Printed White board Production Planning Boards**
  Magnetic whiteboards printed with production

ESTIMATING ▶ ORDER PROCESSING ▶ **PRODUCTION SCHEDULING** ▶ INVENTORY CONTROL ▶ QUALITY CONTROL ▶ COSTING ▶ ACCOUNTING

Figure 4.1 Production scheduling and job planning are an essential element of a streamlined MIS workflow

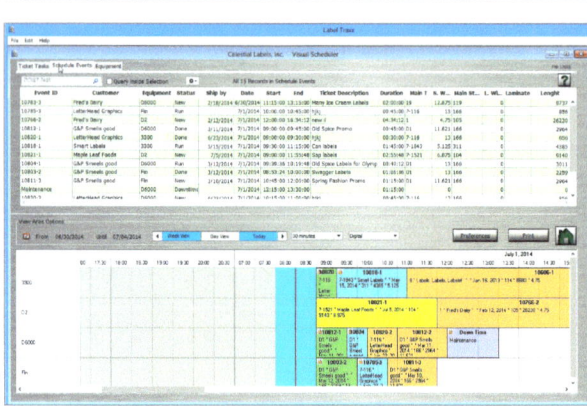

Figure 4.2 Visual scheduler in Label Traxx MIS software system

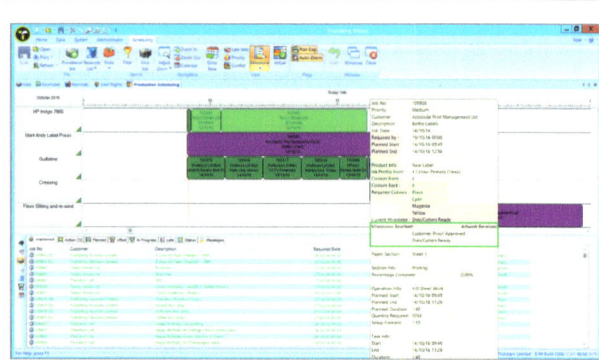

Figure 4.3 Viewing of milestones for scheduled jobs. Source: Tharstern

and project planning grid designs provide an economic planning/scheduling alterative. A dry wipe surface enables data to be easily written and erased.

- **Adhesive Channel Strip Boards**
  An inexpensive solution that uses adhesive backed channel strips that firmly adhere to a backing board so as to hold job card strips, so creating an individual planning/scheduling board.
- **PC with spreadsheet software**
  Spreadsheet software – such as Excel or Lotus 123 – available for use on most computers, can provide an effective means of scheduling using functions such as cutting, copying, inserting, pasting, formatting, coloring text, sorting, and filtering of records so as to provide spreadsheets on which all print jobs can be easily seen on the computer screen (or printed out), re-arranged, sorted and displayed by press or process, order or job number, run length, delivery date, customer or any other requirement.
- **MIS production scheduling modules**
  MIS modules used for production scheduling in the label and package printing plant are available to provide full visibility of machine capacity, both short and long term, and offer detailed scheduling of each production step of the job, from customer approval right through to

shipping (see Figure 4.2). Depending on the software it may also be possible for press operators to consult their work schedule on their press or feedback what they are actually doing. A change in schedule can be automatically seen on the shop floor.

It is also possible to reserve production slots for jobs that are expected, view milestones for scheduled jobs (Figure 4.3) as well as schedule in planned maintenance, down time or holidays. When necessary, a scheduler can move a job up or down the queue, or even split-run the job, part now and part later, or perhaps run in parallel on different presses. A good system can also indicate how much and when overtime will be required. It is also possible to optimize make-ready times by grouping or ganging similar labels or labels with different shapes together, especially if they have the same substrates, colors or cutter tooling.

With some of the more sophisticated MIS systems data can be sent direct to an Esko's Automation Engine for pre-press, proofing and automated step and repeat. More and more digital presses today also come with built-in software to create step and repeat.

Some of the latest computerized production scheduling and control systems used in the printing industry actually look and feel like traditional production board and are very easy to use. Indeed,

Figure 4.4 Screen shot shows an element of scheduling with traffic lights indicating sub-statuses. All need to be green to be able to start production. Source: Cerm

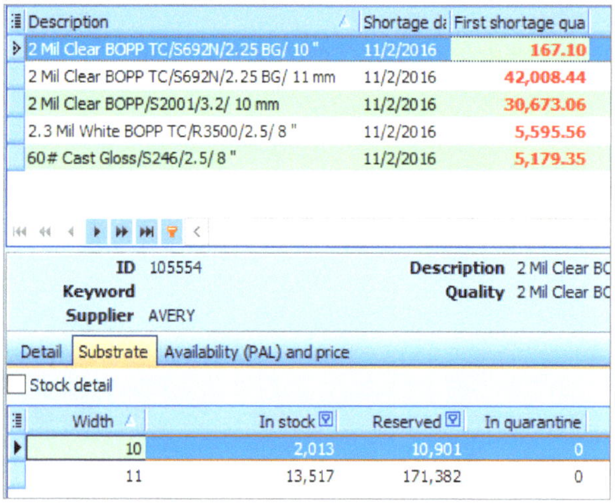

Figure 4.5 A screen shot showing possible materials shortages and alternatives. Source: Cerm

time indicators moving across T-cards as a job progresses can be realistically used to drive a large wall-mounted plasma screen.

It may additionally be possible for a production scheduling and control system to also handle multiple user sites, a wide variety of plant, shift and overtime patterns, schedule multiple component and sections and handle large amounts of digital jobs as easily as conventional analogue printing. Move an item on one screen and it moves on the others. These systems can work standalone or live to the rest of the system by using shop floor data capture and networked machine monitoring.

SCHEDULING AS A CONTROL SYSTEM

Production scheduling, whatever the type of system used, can perhaps be defined as a control system that becomes part of the larger, more-complex label and package printer's manufacturing planning and control system. The overall production scheduling system needs to be considered as more than a schedule-generation process, whether it is manual or automated. It is not just a piece of software, but should ideally be a system that interacts with all departments and provides information that all managers and supervisors need for other planning and supervisory functions.

As can be seen in the traffic light scheduling screen shot in Figure 4.4., all lights need to be green for production to start. Sub status reports can also be shown to provide information on items such as:

- All artwork approved for a job
- Whether there will be enough material in stock on the scheduled production date
- Whether tooling is available

- Whether the order has been confirmed to the customer (to define priorities).

When looking at materials availability, it is important to be aware of possible materials shortages. This can be seen in Figure 4.5, which shows a screen shot indicating material shortages in the next days, based upon current stock, purchase orders and scheduled consumption. The lower part of the screen offers alternative widths and their availability.

It should be born in mind that poor production control and scheduling can undoubtedly be a recipe for disaster. If scheduling is inefficient and not managed properly, then the production floor does not know what to print and when, and machine efficiency can quite quickly fall through the floor. Customers don't get orders filled on time, operators don't get materials when needed, bottlenecks occur – all of which may spell trouble.

Production scheduling is at least as important as any other part of the production loop, and can be the cause of a company's worst nightmare if not handled efficiently and correctly.

A well-ordered and managed planning and scheduling module will undoubtedly provide for and increase the transparency and visibility of a company's

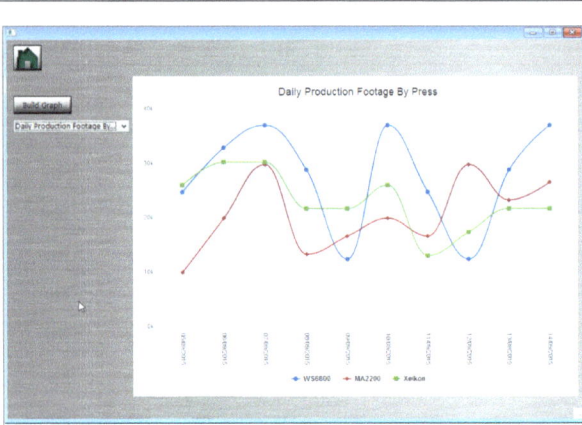

Figure 4.6 Daily production footage by press. Source: Label Traxx

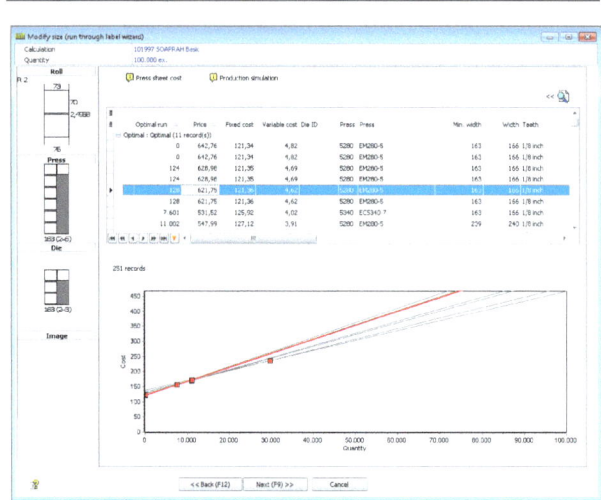

Figure 4.7 Choosing or modifying the best press for the job. Source: Cerm

production plans, help to manage and control progress and capacity right across the business, enabling each department to prioritize and manage their own workloads. Figure 4.6. and Figure 4.7. for example, show how shop floor data collection offers the ability to monitor and track daily press production footage using shop floor data collection.

It is also invaluable in determining whether delivery dates can be met and can identify predictable downtime for maintenance schedules. A proper production schedule also gives production and other personnel a detailed statement of what is expected and in order that supervisors and managers can measure productivity and performance.

Other advantages include minimizing WIP (work in process) inventories, set-up times and overall lead times as well as maximizing machine and production personnel. A good production schedule should also be able to identify resource conflicts, control the release of jobs to the production floor, and ensure that all required raw materials—such as blank product and correct ink colors—are purchased and received on time. Additionally, better coordination will increase overall productivity and minimize operating cost.

Label and package printing production environments may often change dramatically from one day to another, one shift to another, or even hour by hour. This means that the system needs to respond to

the unexpected and quickly be able to identify backlogs and to modify existing production or press schedules (Figure 4.6.). Flexibility in a system is therefore paramount. Keeping it simple makes scheduling manageable. Fast and responsive production scheduling is one of the major keys to successful supply-chain management, in some cases even linking customer's and supplier's schedules together.

Effective and efficient production scheduling today is an essential element in controlling and managing a label or package printing plant, and in providing profitable results.

PRODUCTION SCHEDULING – THE WORK STEPS

To understand how the production scheduling process works within an MIS system, Cerm have kindly supplied the following screen grab (Figure 4.8) to explain the necessary work steps involved. As can be seen, there are four columns on the screen – Description, Status, Scheduled and Waiting time.

- The Description column shows the job order details, followed by the order date, proof date, plates, press, rewinding, finishing, shipping and delivery.

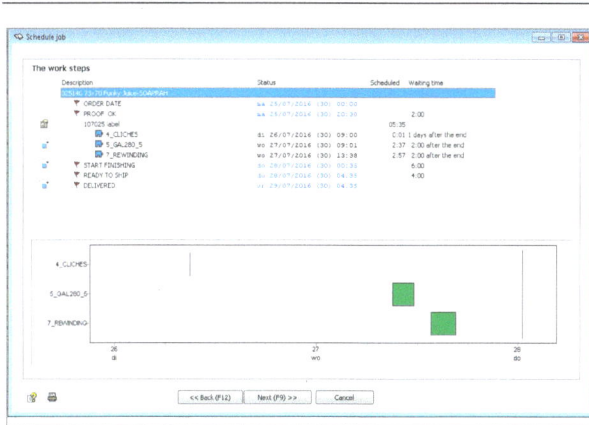

Figure 4.8 Work steps involved in production scheduling using Cerm MIS

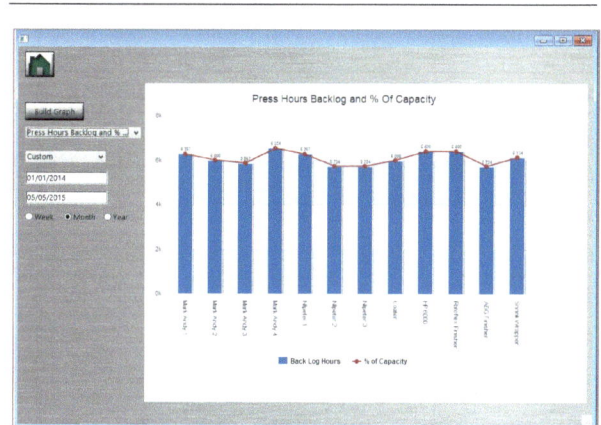

Figure 4.9 Press hours backlog and % capacity. Source: Label Traxx

- In the Scheduled column, every work step is allocated a calculated time.
- In the Waiting time column, every work step is allocated a waiting time (queuing time needed to go to start the next step).
- Based upon delivery date and transport time, the MIS system calculates backwards (Status column) to determine the start of every step necessary: rewinding, printing, plate-making (depending on the complexity of the job) the allocation of resources and the sequencing of timing and tasks necessary to produce the printed products and As indicated in the illustration, MIS production scheduling is an automated system for dealing with related services.

EASIER DECISION MAKING

In summary, production scheduling using one of the latest MIS systems will make the key work steps in label and package printing decision making easier and simpler, in particular:

- In releasing jobs for production
- By prioritizing jobs that require resources (spot colors, special tooling, stamping foils, and so on)
- By assigning resources (people, presses, inspection machines)
- In reassigning resources from one job to another

(similar jobs)
- By determining when jobs should be started (to meet deadlines/delivery dates)
- By halting a job that can be stopped to make way for another more urgent job (a rush order).

Quite simply, using a good MIS production scheduling system will provide label and package printers with faster production, lower costs, and increased accuracy and should offer:

- Ease and simplicity of use with a similar look and feel to a manual scheduling board
- Simultaneous scheduling of all resources
- The ability to provide overdue, overloaded and other critical path warnings
- Good activity tracking and the ability to view availability of paper, plates, dies, etc.
- An effective 'what if' scenario to analyze the impact of equipment or business changes
- The ability to plan human resource requirements
- A full reporting facility including press hours backlog, available capacity (see Figure 4.9), hours, job run lists and actual production performance.

Today, a sophisticated MIS system with efficient job planning and production scheduling capabilities is more than a nice-to-have tool; it is becoming a necessity.

Chapter 5

Inventory control – raw materials, warehouse and finished goods

Most label converters and package printers carry a fair degree of stock from paper and film reels, to sheet materials, inks and varnishes, cutting, embossing and foiling dies, spare parts, hot and cold foils, cartons, tapes and staples, and other packaging materials, as well as holding stocks of finished products for customer call-off, or even pre-printed stock products labels or packaging. New label or printed packaging stocks, including inks, cutters, other tooling, packing and cleaning materials, foils, lamps, etc. are coming in all the time.

All these stocks of production items or finished goods need to be managed as efficiently as possible, and this is undertaken in an MIS inventory management or stock control system – a set of hardware and software based tools that automate the process of tracking inventory. Inventory management and stock control may also be closely integrated with 'Purchasing.' The kinds of inventory tracked with an inventory control system can include almost any type of quantifiable goods that are likely to be used in a label or package printing plant.

Controlling and managing both stocked raw materials awaiting production and finished goods inventory awaiting shipping, as well as warehouse tracking, is undoubtedly a key part of any label or package printing management information system – ideally tracking product inventory in real time by scanning a barcode on the item to be recorded and tracked. Some inventory control systems are able to work in real-time using wireless technology to transmit information to a central computer system as transactions occur. The aim should be to know what materials or goods are on-hand, what items are on order, and what items are needed. Done efficiently, this will minimize flor or storage stock.

When integrated with purchasing and order entry (Figure 5.1), stored materials may be automatically allocated to a particular job. As incoming goods are received, an easy-to-use search list identifies them against purchase orders. Each incoming pallet, roll, plate, tool or customer stock can then be tracked into and out of storage by a serial number or bar code, with the quantity received auto-allocated to each job or storage position.

With such sophisticated systems it also becomes feasible to agree and store both minimum and maximum stock holdings, avoid unnecessary inventory purchase and production bottlenecks, with the system providing replenishment reporting and even automated inventory management. Inventory control can therefore range from relatively simple to quite complex systems. However, they will most commonly incorporate the same basic key areas of inventory and stock control management.

These three key areas of inventory and stock control that need to be integrated within an MIS will need to cover raw materials stock holding, warehouse management and tracking, and finished goods inventory. This is shown as a flow chart in the Fig 5.2.

Quite simply the aim of an efficient inventory and stock management control system or module should be to ensure that materials are always available to fulfill orders, to show how much stock is held at any one time, to know where it is, and to make sure that jobs are not delayed in production. The system should also ensure that capital is not tied up in surplus or old stock, thereby reducing wastage and optimizing stock levels, as well as offering full traceability and reporting as required. So let's amplify the three key areas of inventory and stock control in rather more detail.

RAW MATERIALS STOCK CONTROL

A good raw materials stock control system will enable the user to pre-allocate roll material or sheet stock to jobs, show current stock balances and values, together with providing an extensive audit and location trail – job allocation, issue time/date, movement, receipt – of all transactions from the final end job back to its original supplier source, whatever press or process it has been through. Ideally, it will also show traceability of all other jobs produced from

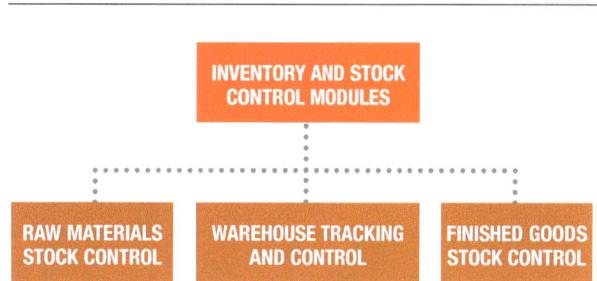

Figure 5.2 Inventory and stock control modules integrated within MIS

the same batch of material.

The stock control system should integrate seamlessly with the order processing, production scheduling and management, quality control and job costing modules. Each action within each of these modules will automatically update related stock levels, keeping stock figures accurate and up to date without the need for manual data entry

Processing an order for production automatically 'allocates' the material required in stock control, while raising a purchase order notifies stock that the material is 'on order', and recording the materials used through job costing or remote data capture removes it from stock.

Most of the MIS systems for labels have a roll tracking facility where rolls are barcoded and through job costing get scanned against a job (see Figure 5.3). This enables full traceability of the rolls used on that job. Some MIS are integrated with a number of material suppliers for electronic ordering with a PO sent electronically. The shipment details then get sent back to the MIS electronically, with the manufacturer's

Figure 5.1 Inventory control should be a key element in a streamlined MIS workflow

Figure 5.3 Barcode scanning of rolls using an iPod scanner. Source: Label Traxx

Figure 5.4 WiFi barcode scanner. Source: Cerm

EPSMA barcode number of the roll and the exact width and length and, depending on the MIS supplier, information about the adhesive, caliper, coatings, and suggested stock substitutions. An iPod scanner can then be used to take roll inventory by scanning all of the roll barcodes, and even allocate specific rolls to certain jobs.

Codes, including barcodes (see the Label Academy book on Codes and Coding Technology for suitable codes), can undoubtedly make the whole process of inventory control and stocktaking much easier, but it can still be quite time consuming. Checking stock more frequently – a rolling stock-take – avoids a massive annual exercise, but demands constant attention throughout the year.

For more sophisticated inventory control then Radio Frequency Identification (RFID) tagging using hand-held wireless (WiFi) barcode readers can offer a simple and efficient way to maintain a continuous check on inventory in real-time on the shop-floor. Whatever the system, the aim should be to ensure 100% traceability of all materials.

For label and packaging converters working with environmental management control or environment audit systems, then their MIS inventory control may also need to add and provide for materials 'certification' details (such as FSC certification) to be entered and displayed as required by the leading environmental agencies.

Depending on the warehouse and materials storage process being used, its size, location and management, the MIS may need to incorporate a stock search and viewing facility that will easily identify stock in a particular location (Figure 5.5). This may in turn mean that a definable storage structure that allocates stock to aisles, rows, levels, etc., has to be established if not already in operation.

On first-in-first-out storage principles the system will show current stock balances and values with an audit trail of all transactions. There should also be full batch traceability from an end product back to its source via whatever machines it was produced on. It should also show the history of all other jobs produced from that same original batch.

A reporting facility will enable ready stock management and should be able to facilitate periodic or annual stock taking, as well as ensuring that stock is not overlooked, that the right balance of stock is in the right place at the right time, that jobs are not held-up, and that money is not tied up with surplus or old outdated stock. It should also be closely integrated with estimating, purchasing, costing, production control, quality control and waste management and ideally be based upon the requirements of ISO9000.

Achieving ISO or BSI quality standards is

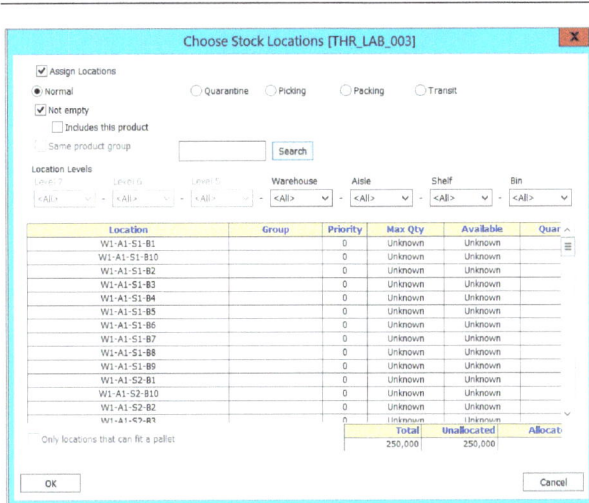

Figure 5.5 Stock can be assigned to locations in this Tharstern software

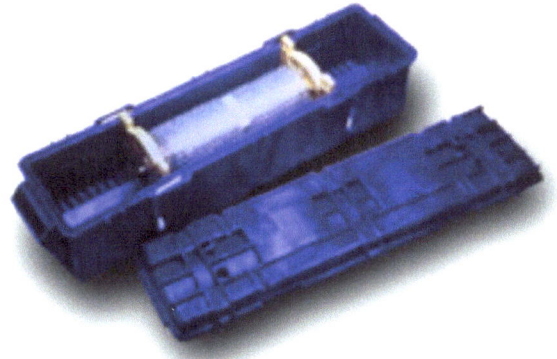

Figure 5.6 Use manufacturer's shipping boxes for safe tooling storage. Source: Rotometrics

undoubtedly one way of showing customers and regulators that you take quality control seriously. Quality control can therefore be a vital aspect of stock control - especially as it may the quality of the finished product.

Efficient stock control should incorporate stock tracking and batch tracking. This means being able to trace a particular item backwards or forwards from source to finished product, and identifying the other items in the batch. With built in reconciliations for physical inventory adjustments, purchase orders, material receipts, damage roll tracking, spoilage tracking, and shipment cost recording, converters can be confident that the appropriate costs are being recorded and more importantly, billed.

Goods should be checked systematically for quality so faults can be identified and the affected batch or product weeded out. This will allow the converter to raise any problems with the supplier and demonstrate the quality of the products produced. Some systems, such as the Tharstern Stock Control module are also able to enable the addition of 'product certification' details to materials, such as FSC certification for environmental audit trail and analysis

Remember however, that any inventory control system – whether manual or a sophisticated fully

computerized system – is only going to be as good as the original and ongoing data that is entered into it.

STORAGE OF DIES AND CYLINDERS

The creation and maintaining of a tooling library of cutting, embossing and foiling dies, cylinders and specialized tooling is an important element of stock management and control. The manufacture and use of tooling at the finishing end of a roll-label press often involves engineered products that can be heavier, harder to handle, bulkier and potentially more likely to be damaged or cause damage than in many other label productions applications. Quite simply, moving and handling precision engineered tools during manufacturing or in warehouse, production or storage areas requires specific expertise and training. In particular:

- Solid rotary cutting, embossing or foiling tools, as well as cylinders and anvils, can be quite heavy and a challenge to lift and move in or out of the storage area and handled safely without risk to the tool or the handler
- Flatbed, rotary or flexible cutting dies have very sharp edges. Handlers can sustain cuts and the cutting edges can be damaged during handling and preparation for storage
- Unpacking of incoming cutting dies and other

tooling – and re-packing for storage or shipping – again has the potential to cause damage to the tool or operator during handling

- Poor tooling maintenance and storage conditions can lead to deterioration in tools over time.

Such considerations mean that the systems that are used to move and handle tooling as it enters from the supplier, is stored prior to production, moves from incoming goods storage, passes through, and departs the press room, and then back to a storage area for cleaning, packing for storage, and then soring for easy and quick location can be critical to a company's productivity

It is not just a case of putting tooling into storage after use. They need to be cleaned, treated (oiled) prior to storage, inspected, protected with chemically treated wrapping, or placed back into the manufacturers shipping boxes (see Figure 5.6) for safe storage when possible.

Good storage, transportation and handling systems for all tooling used in label converting operations can significantly reduce costs, increase productivity and create a safer, more ergonomic production environment. Key factors that therefore need to be considered and addressed in the handling and storage of tooling in the label production plant are:

- Checking that there is no transportation or handling damage, either from or to the manufacturer or within the storage facility
- The elimination as far as possible of any form of tool damage
- Making use of optimum handling and storage procedures
- The recording of the exact storage position of every item of tooling
- The provision of easy access to each individual tool in storage
- The elimination or minimizing of any heavy or awkward lifting operations
- Obtaining a significant decrease in the chance of accidents.

As can be seen, the storage, handling and identification of tooling from incoming tools, during storage, for easy production location in store, after

Figure 5.7 Mobile menu on a WiFi handheld warehouse scanner

production storage, and then possibly returned to the supplier for repair or sharpening, can be a key to employee safety, productivity and profitability. A raw materials MIS system that can successfully and accurately provide for all these stages – depending on the volume of tooling, the range and variety of tools used, the availability and size of storage area – needs to be carefully discussed with the MIS supplier.

If label and packaging printing plants use external manufacturers for the supply of printing plates, then the same kinds of issues of handling, checking, cleaning, packing and storage of tooling are also going to occur. Again, discussion with the MIS supplier to create a suitable system for the plant needs to be undertaken.

WAREHOUSE CONTROL AND TRACEABILITY
Relying on a paper trail and the manual entry of data to manage a label or package printing warehouse is not a very satisfactory means of achieving worker productivity and inventory accuracy. Once received as

Fig. 5.8 A hand-held scanner being used to read barcodes in a warehouse

Figure 5.9 Printed closing (left) and core labels (right) for roll identification. Source: Cerm

a computer print-out or even a hand-written order, pre-production stock or perhaps finished goods that need to be packed and shipped, will be picked from the warehouse, with a paper trail tracking every step of the process. The information is then manually entered it into the system and filed. Should a discrepancy appear in a customer's order or invoice, pinpointing the problem requires cross referencing the data in both the system and in filing cabinets.

On the other hand, a good mobile warehouse WiFi scanner with a menu-controlled tracking system (Figure 5.7) within an MIS workflow will automate the flow of information and coordinate key activities in the warehouse in order to maximize efficiency and undoubtedly help to streamline a label or package printing business, enabling the warehouse to be configured into easily identifiable areas and for the rapid tracking and movement of incoming stock, stored stock, finished stock and any work-in-progress.

It should also show goods that have been delivered, but have not yet been allocated a final location, as well as knowing the location of all existing inventory, being able to direct workers to the right storage location(s), minimize the order picking process, and reduce any impact from products out-of-stock. Get all of these working well and errors will be eliminated and productivity will be improved.

Depending on the MIS warehouse or inventory system or software, it may only be suitable for a single site or be suitable for handling multiple sites. Sites can be configured for tracking and traceability in a variety of ways, from very simple areas, white paper or reel stock warehouse, machine room, despatch, etc. to very detailed layouts, warehouse, area, bay, rack, shelf, etc.

To make it easy for warehousing systems to record and track reels, pallets, sheets, inks, etc., it may be viable to install shop floor data capture stations or use mobile hand-held scanners (Figure 5.8) or wireless palm computers. Barcoded reel, pallet or box labels can be generated within such systems, enabling users to locate stock items very quickly by a range of different search criteria, including stock codes, job number, department or section, picking note.

Installing a full-scale warehouse management system can be quite a costly investment, yet its value is usually quickly recognized. After a warehouse management system has been up and running for only a few months, users tend to say that they cannot envisage how their business worked without it. From achieving real-time visibility into inventory and orders, to decreasing the time it takes to invoice and receive customer payments, a warehouse management information system and package extends mobility to warehouse workers and prepares the business for further expansion.

FINISHED GOODS AND STOCK PRODUCTS CONTROL

Finished goods and stock products inventory includes

barcoded core labels (Figure 5.9) and printed packaging called-off by customers as well as things that are bought to be resold like ribbons and printers. There should also be a means of setting expiry times on stock. Increasingly, MIS suppliers are also being asked to incorporate interfaces to transport suppliers, enabling the sending of electronic instructions to DHL, UPS, etc., in order to manage transport and shipping.

The primary function of finished goods and stock products control is to manage all printed stock and consignment stock held either on behalf of customers or for direct re-sale – including managing inventory cost. This all needs to be closely integrated with works instructions, delivery notes and invoicing, provides a fast and seamless stock movement and management system that greatly reduces the opportunity for human error.

All the features of raw materials stock control will also apply to finished good control; first-in-first-out full audit trail; batch control and barcode traceability; a sort-filter-find engine and full integration with report and document generation.

Finished Stock call-offs can be received and entered either manually or through EDI and e-Commerce web-to-print orders to generate picking notes.

Chapter 6

Quality control and compliance management

The printing of labels and packaging involves much more than just putting ink on paper, board, film or metallic foils. Yes, jobs may be printed in anything up to eight or more colors – from flexo to offset, screen, photogravure, electrophotographic or inkjet – like some other printing sectors, but this can be followed by a wide or diverse range of converting or finishing line operations. These operations may include hot or cold foiling, embossing, die-cutting, varnishing, laminating, scoring, perforating, folding, re-winding or sheeting.

To perhaps make things even more complex there are an increasing number of jobs that may be printed on multi-process combination or hybrid presses, using two, three or more different printing processes in one production line. Indeed, hybrid digital/analogue presses are forecast as one of the growth technologies for the future.

The more diverse the range of substrates, the more printing colors and processes used on a job, the more converting and finishing operations, then the more chance there is of something going wrong or of quality being compromized. Possible problems can range from:

- Poor register
- Color variation
- Picking
- Hickeys
- Poor die-cutting
- Poor transfer of foils
- Waste stripping problems
- Missing labels
- Poor slitting or scoring
- Bar code verification problems
- Adhesion problems
- Dispensing and application problems

The list could go on, but what can already be seen is that the range of things that can cause quality, accreditation or compliance problems in label and package printing is quite diverse. The more quality, accreditation or compliance variations or problems that a buyer can find, the more chance there is of the job or part of the job being rejected or a discount negotiated. It may even lead to the loss of accreditations or the supply contract.

Little wonder then, that the industry increasingly makes use of technology and systems that can pick-up print register or color problems, identify faults, verify barcodes, as well as check or inspect a whole range of other quality or possible standards variations. In addition, most of the leading brand owners and retail group will expect their label and pack suppliers to have some kind of quality control, quality management, quality audit, statistical process control or compliance management system which, today, is increasingly being integrated into a sophisticated

Management Information System (MIS) and automated workflow process. See Figure 6.1.

Having a system that provides a platform for quality, accreditation or compliance control, with detailed written procedures, recording and documentation of quality and performance issues, employee training, returns procedures and complaints logging has therefore become an important element in the management of label and package printing production information.

When problems or complaints arise there also needs to be a system of reporting and corrective actions, including the generation of management reports that enable faults or compliance trends to be ascertained and analyzed, perhaps leading to updating of quality control or compliance procedures or new employee quality training documentation.

Quality control systems may also store ISO, OSHA or other international or national Standards requirements and documentation that can be viewed or print generated by employees. Depending on the end-user sector, there may also be specific food, pharma, industrial regulations or compliance requirements. So what quality control, quality management, quality standards, compliance or accreditation schemes may label and package printers be working to?

Quality Control, also called Quality Management, Statistical Process Control (SPC) or Total Quality Management (TQM), are all management systems in which tests and/or inspections are applied at various stages during inventory management, during label or package printing production and during finishing to ensure that the end product meets pre-determined specifications and standards of quality and performance.

For the printers of pharmaceutical or food labels and packaging, there are also other quality or performance systems that they may have to work to: FDA- or EU-compliant materials, Good manufacturing Practice (GMP) or Certificate of Compliance (CoC) for pharmaceutical products and medicines, and BRC Global Standards for the food and retail sector suppliers.

In the label and package printing industry, ISO 9000 has become the standard by which key label users, brand owners and buyers assess whether a system and procedures are in place to check and test incoming materials, or manage and control quality. More sophisticated quality management and quality control systems now being adopted by the label and package printing industry include Statistical Process Control, Total Quality Management and Six Sigma. See Figure 6.2.

All these quality systems and standards outlined in the flow chart, together with GMP, CoC and BRC Global Standards, are amplified in rather more detail as follows:

ISO 9000. Achieving certification to the internationally recognized quality management standard ISO 9000 has become the most common way for the label and package printing industry to introduce and implement a quality management system and demonstrate its commitment to quality. For many label and packaging industry suppliers and converters – as well as label buyers – it has become the expected standard.

Implementation of the BS EN ISO 9000 quality standard involves making a large number of small quality improvements across a wide range of business processes. Activities can be managed as a process, with each activity adding value to the last, all ultimately leading to customer satisfaction. Simple, user friendly, documentation can be created around a

Figure 6.1 Quality control is becoming an integral part of an MIS and automated workflow process

Quality management and control systems			
BS EN ISO 9000	**SPC Statistical Process Control**	**TQM Total Quality Management**	**Six Sigma**
International standard to introduce and implement a quality management system and demonstrate commitment to quality	Formerly called Quality Control, SPS depends on Quality Assurance functions of raw materials control, print process control and tolerences, and on inspections	A management philosophy in which all levels of employee participate in continuously improving the quality of products and services. Successful management of quality performance	Similar to TQM but more quantitative and more pragmatic in its approach, using statistical analysis performance management

Figure 6.2 Quality management and control systems used in the label and package printing industry

company's existing business practices.

Statistical Process Control (SPC). A system by which printed label and pack quality results can be controlled, Statistical Process Control (more commonly known as Quality Control in the past), depends on the Quality Assurance functions of raw materials specifications and controls, printing process control, on acceptable standards and tolerances and on inspection. Each of these elements is further outlined below. Also Figure 6.3.

- **Specification and control of materials.** Label printing uses a wide range of raw materials – label stock, ink, adhesives, varnishes, blankets, rollers, cutting dies, etc. – which should all ideally be controlled. Raw materials properties that can be measured, tested and controlled include the moisture content, brightness, gloss, absorption or picking of paper face materials; the tack, gloss, drying, color, yield or flow of inks or varnishes and Shore hardness of blankets, plates or rollers.
- **Printing process control.** The control of printing and processes can begin at the press fingerprinting stage, while color bars, targets, grayscales and other devices or instruments are used to provide objective measurements and

numbers to control production. Instruments used for checking at various stages throughout the color origination, proofing, or printing process stages include spectro-photometers and densitometers.

- **Standards and tolerances.** Agreed standards and tolerances should preferably be established between the printer and the customer at the time the job specifications are finalized and accepted. These should aim to agree on tolerances for color variations, density variations, register variations, cutter tolerances, etc. and will vary according to the quality of the labels being produced.
- **Inspection.** The use of inspection procedures is essential in ensuring that SPC systems function properly and that materials control, process control and standards and tolerances are all functioning properly, ensuring that quality is improved and waste reduced.

Total Quality Management (TQM). The basis of TQM is that any product or service can be improved upon and that improvements reduce costs, provide better performance and give higher reliability. The approach of TQM is that any company can

```
                    ┌─────────────────────────────────┐
                    │   Statistical Process Control   │
                    └─────────────────────────────────┘
```

| Specification and control of materials | Standards and tolerances | Printing process control | Inspection |

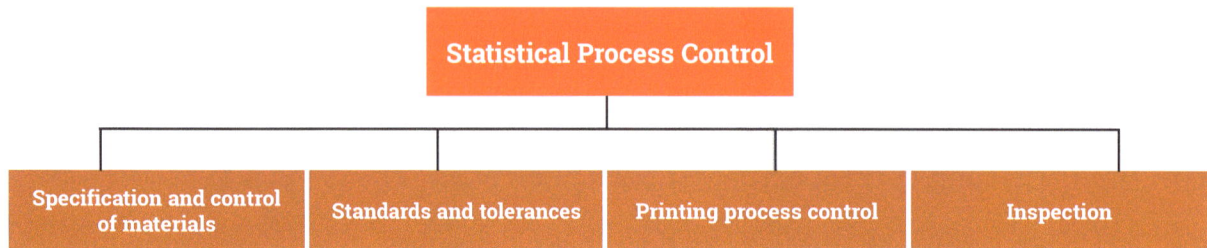

Figure 6.3 Key elements in the operation of a statistical process control system

continuously improve the quality of its goods and services – manage its quality performance – through the participation of all levels of employees.

Developed in Japan after the Second World War, TQM's management approach is aimed at long-term success through customer satisfaction. It is based on the participation of all employees within a company or group and on improving processes, products and services. Key elements within a TQM program include:

- The customer defines quality. His requirements are confirmed and any problems identified
- Senior management is responsible for taking the 'quality' lead in developing the company's quality culture and strategy. All employees should be involved
- Quality is dependent on design and execution of all processes and systems to a high standard. It should be designed in from the beginning
- Error prevention should be built in to work processes and systems
- The 'cost of quality' or the 'price of poor quality or non-performance' should be calculated
- Problem solving techniques should be developed and implemented
- Continuous quality improvement is key to achieving higher quality standards
- Suppliers should be involved in the achievement of quality targets
- An ongoing aim should be to shorten response times
- Results should be communicated to employees

Label companies already working to ISO 9000 standards can move to higher levels of quality management through implementation of TQM or Six Sigma.

Six Sigma. Six Sigma is a management methodology that enables businesses to achieve a goal of increasing profits, improving product quality and enhancing employee morale by eliminating waste, decreasing process variation and by identifying and reducing defects in manufacturing. It uses information and statistical analysis to improve a company's operational performances, practices and systems by identifying and preventing 'defects' in manufacturing and related processes, enabling companies to both anticipate and exceed expectations.

More than just a quality system like ISO or TQM (Total Quality Management), Six Sigma is more about a philosophy or way of doing business in which defects are measured per million opportunities. Developed by Motorola, the Six Sigma methodology can be defined by reference to three levels: metric, methodology and philosophy.

- Metric. Commonly referred to as 3.4 Defects Per Million Opportunities (DPMO) provides a guide for considering at least three opportunities for a physical component.
- Methodology. The tools and roadmap that provide a structured method of solving problems.
- Philosophy. The route to reducing variation in a business by taking customer-focused, data driven decisions.

Six Sigma and TQM have many characteristics in common. They both focus on quality control, both address quality problems at their roots, both focus on solutions geared to the overall goal of the business, both rely on data as evidence, both require senior management to buy into the quality program and both aim for a lasting cultural change throughout the whole organization. Six Sigma however, is more quantitative and more pragmatic in its approach, using statistical analysis performance management. Six Sigma can be incorporated within an existing TQM environment.

GOOD MANUFACTURING PRACTICE (GMP)

A GMP is a system that is used for ensuring that products are consistently produced and controlled according to quality standards. It is designed to minimize the risks involved in any pharmaceutical production that cannot be eliminated through testing the final product.

GMP covers all aspects of production from the initial starting materials, premises and equipment to the training and personal hygiene of staff. Detailed, written procedures are essential for each process that could affect the quality of the finished product. There must be systems to provide documented proof that correct procedures are consistently followed at each step in the manufacturing process - every time a product is made.

CERTIFICATE OF COMPLIANCE (COC)

A certificate of compliance (CoC) with the European Pharmacopoeia (also called CEP) is a document published by the EDQM (European Directorate for the quality of medicines) and which certifies that a substance with pharmaceutical purpose is well manufactured and controlled according to the equivalent monograph of the European Pharmacopoeia.

For a CEP, the manufacturer of a substance must file with the EDQM a technical file describing the product, its manufacturing, controls effectués… mode. It is assessed by members of the EDQM and the manufacturing site is inspected by one of the European agencies.

The advantage of the CEP is considered that the technical file of the substance was assessed and that

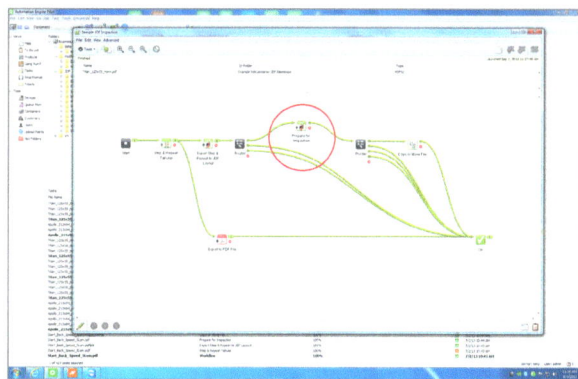

Figure 6.4 JDF job inspection set-up in an AVT system

on applications for authorizations to market multiple of a drug in Europe, it is more necessary that each national agency to do this assessment. Thus, CEP represents a kind of guarantee of quality and is therefore heavily used to active principles, especially if they are produced in some countries (such as the India or China) where quality standards differ from those in Europe.

In addition, by 2019 more than 80 per cent of global drug production will need to be complaint with regulations regarding unit identification.

BRC GLOBAL STANDARDS

BRC Global Standards is a leading safety and quality certification programme, used by over 23,000 certificated suppliers in 123 countries, with certification issued through a worldwide network of accredited certification bodies.

The Standards guarantee the standardization of quality, safety and operational criteria and ensure that manufacturers fulfil their legal obligations and provide protection for the end consumer. BRC Global Standards are now often a fundamental requirement of leading retailers.

MIS QUALITY CONTROL SYSTEMS

Starting with ISO 9000 as an internationally recognized quality management standard, it is possible to bolt on any or all of statistical process control, total quality management and Six Sigma systems to MIS systems to eventually rise to the highest levels of quality management and customer quality satisfaction.

Today, some of the leading MIS systems for label and package printing incorporate quality control modules and are able to integrate with Esko's Automation Engine, as well as to presses, AVT inspection (see Figure 6.4) and ABG finishing equipment to automatically collect and analyze quality control data.

Esko's Automation Engine serves as the heart of any size prepress production operation, enabling unparalleled workflow automation with rock-solid quality control. It comes with extensive business system integration capabilities and is highly scalable.

A job in Automation Engine represents a production order that organizes the data storage for the job but also the job's metadata, its link to order ID, due date, customer info, Customer Service Representative contacts, etc.

Besides these administration attributes, a job can also contain the graphical specifications like barcode, inks, RIP options. This job information can be used in any workflow to take advantage of all the data that is already there, avoiding double entries.

Importantly, Automation Engine ensures increased efficiency and throughput and saves time and money. It is, by all standards, an exceptional answer to the daily challenges of print professionals who need to increase quality, reduce errors and drive cost out of the process. It is also able to obtain information from external systems through SQL queries, XML or XMP data. This data populates job information and workflow parameters, using values that already exist. Examples include step and repeat data, barcodes, colors to be used, etc.

All job information is kept in an internal database, which can be archived for future use. Order administration systems can be linked through an MIS integration project, such as those undertaken by Cerm, Label Traxx, Tharstern, EFI, Global Vision and other suppliers. Information and data are centralized, which reduces operator intervention. Savings are achieved by measuring and billing consumables, proofs and plates.

The following paragraphs look at how some of these leading suppliers enhance Quality Control and Quality Assurance.

With the **Label Traxx Quality Control Module**, it is designed to assist in improving shop-wide quality control and the documentation of quality issues and costs. The Module incorporates five sections: Quality Procedures, Return Materials, Non-conforming materials, Complaints log and Documentation.

Put together these sections set out to improve shop-wide quality through the use of recommended quality procedures – embedded into the workflow – and configuring the interface for the capture storage, and display of recommended procedures for all production events. It will also capture and display customer-specific procedures. Importantly, it eliminates all of the Post-It notes, notebook and high-light forms that employees often forget to look at.

The recording of product returns with reason codes and corrective actions for reporting processes is also available within the Module, as well as the recording of non-conforming materials from vendors. It will also store and present OSHA and ISO requirements systematically.

Taking a different approach to Quality Control, **Cerm** consider 'Quality' as inherent to daily practice. Starting with quality at a GMP-certified customers first, they found that 'quality is a work attitude,' and not just a checklist. The most important Quality Assurance features available in Cerm Software include:

- Approval procedures
- Project related screens
- Quality level management
- Quality groups
- Checklists
- Shopfloor line clearance
- Unique identification of raw materials and finished goods
- Sampling
- Traceability
- Registration and the following up of complaints.

An unlimited number of quality levels can be created in the software, containing rules for product approval, product ganging, product mixing and batch release. These can be set as a default per customer. It is also possible to impose an approval cycle for every new calculation, product, sales order and job, and

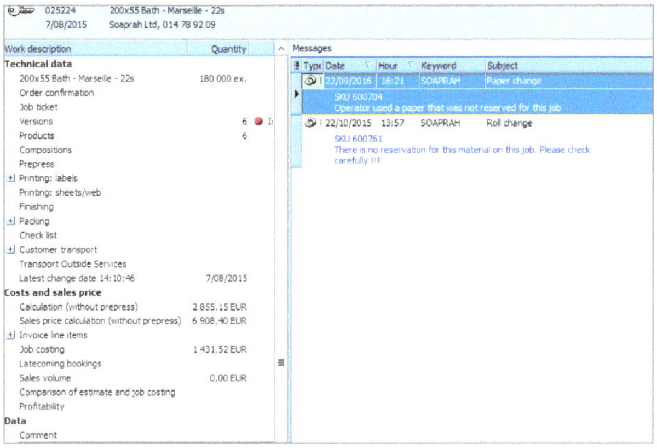

Figure 6.5 The Cerm complaints software displays messages about problems or complaints

these are created in the software.

Based on a database of of quality checks the series of checks that need to be executed – based on certain criteria (multiple appearance conditions) – are defined. This forms the basis for the printing of a Quality checklist report per job. The document is to be filled out by operators and QC/QA throughout the flow.

For complaints, Cerm have developed what they call a 'message' module, which stores all communications – visit reports, customer communications, supplier communications, as well as 'complaints', production remarks, etc. – that should be kept in a central database, although most of the time hidden. These messages can be found with an easy 'click', and can contain questions about the cause of complaints (see Figure 6.5) and descriptions about the solution.

Cerm software is fully JDF/JMF compliant and integrates this functionality in the complete production flow. A JDF workflow reduces risks and mistakes by avoiding data re-entry and automating manual, repetitive steps from the production process. Therefore links exist with pre-press workflows (already mentioned above with the Esko Automation Engine), Printing-machines, Finishing-machines and Camera inspection systems (see AVT below).

Another MIS vendor with a Quality Management software solution is *EFI*. By setting up quality tests for

specific raw materials, vendors, finished goods or customers, the EFI software can ensure that quality materials are being placed into inventory and that customers receive the products they expect. As quality issues arise, corrective action reports can be generated. And, through integration with DMI/Shop Floor Data Collection and scheduling, rework and schedule changes become less disruptive.

The latest version 4.1 of the EFI Packaging Suite, introduced at Labelexpo includes certified workflows geared toward real-world label and package converting environments. Each certified workflow combines the core EFI Radius ERP software with modular, integrated components from EFI's portfolio, as well as with key third-party technologies, such as Esko's Automation Engine.

One of the components in the Suite, the new Auto-Count 4D shop floor production intelligence platform, is making its worldwide debut. This innovative new component automatically collects accurate, up-to-the minute production data including counts, press status, speed and other critical information directly from production equipment in real-time. Through full plant visibility and data-driven reporting now in a browser based environment, packaging printers can have all of the information they need to make decisions in seconds.

Within the *Tharstern* Job Tracking Module there is the ability in their print 'passport' to track and record completion of required quality procedures during operations and ensure departmental compliance with quality assurance procedures e.g. operator confirms the imposition has been checked and approved. Operators can additionally add and edit their own notes on a job; these can even be emailed to a designated address. Milestone restrictions prevent jobs from being started if certain milestones have not been achieved, such as plates and material not ready.

OTHER QUALITY MANAGEMENT AND SOFTWARE SOLUTIONS

Outside of the main label and package printing MIS system, there are now an increasing number of print quality inspection, color, fault, barcode and other inspection solution available, both software and

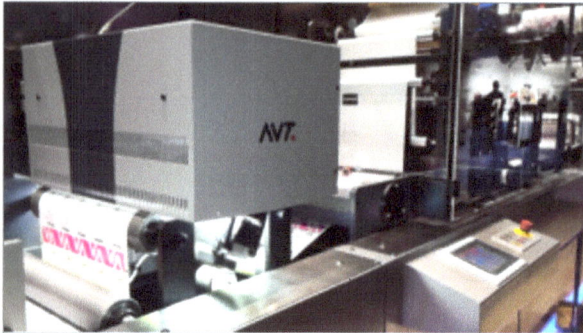

Figure 6.6 Illustration shows the AVT Helios Turbo HD for high definition, high resolution quality inspection

hardware – many of which are already being integrated with Esko's Automation Engine and MIS systems. Some of the key solutions are briefly described below.

AVT recently premiered their Helios Turbo HD, which offers high definition resolution inspection along and across the web at full press and rewinder speeds – without sacrificing quality assurance and key applications such as in-line barcode verification.

To meet the specific needs of digital printing in the label and narrow web markets AVT introduced Helios D, an automatic, 100 per cent print inspection system. Based on AVT's Helios product line, Helios D supports all stages of digital production workflow, including the identification of specific defects like missing nozzles, ink dripping spots, and color changes as soon as they occur. The result is reduced waste, enhanced production and comprehensive process monitoring, according to the company.

A new generation of AVT's in-line spectral measurement supporting nearly any application, including transparent flexible, paper, cartons, etc., SpectraLab II offers a compact design, advanced color workflow management and improved in-line to off-line measurement correlation. SpectraLab II also features enhanced color reporting capabilities, including real-time alerts to third-party reporting systems when colors shift out of tolerance.

The AVT systems extract business intelligence from the production floor and provide seamless connectivity to MIS and pre-press for optimized automated workflows.

Like some other quality control equipment and systems suppliers **Global Vision** has also embedded a Quality Control Platform within Esko's Automation Engine so as to provide an all-in-one quality assurance solution that runs as a background process and results in an annotated and viewable design file within Automation Engine. This further enables rapid and systematic review of all detected packaging and labeling errors throughout the workflow, from upstream design to print.

Being able to process more jobs efficiently, while reducing the inherent risk of errors is critically important for label and package printer customers. By integrating Global Vision's Quality Control Platform into Automation Engine, it becomes possible to address the growing complexity of artwork quality assurance within a single solution that connects to other Esko applications in Esko Software Platform 16.

This approach enables everyone from graphic and print suppliers to brand owners to have access to automated quality control tools. Automated quality checking tools additionally enable label and package printers to process more jobs with the same people and yet with fewer errors. Being able to process more jobs efficiently, while reducing the inherent risk of errors is critically important today.

A 'unique, highly adaptive' new color management software solution for label and packaging applications is available from **QuadTech**. Without any hardware modifications to the press, QuadTech's ColorTrack software integrates with ink formulation software to simplify workflow and reduce the number of ink corrections needed to achieve accurate, optimal color.

Meeting the demand for more in-line finishing and shorter runs has become easier with such integrated systems. Operators can control the entire production process from reel to reel. Miniature scanning heads recognize everything from standard colors to inks and coatings, invisible to the human eye, reducing job set-up time and waste.

It was because meeting the quality and consistency demands of brand owners had become increasingly difficult, that QuadTech, in partnership with ColorConsulting, developed what they describe as a 'color expert in a box' that automates the process of

Figure 6.7 QuadTech Color Measurement Screen showing trends

Figure 6.8 Taken from a QuadTech Inspection screen

delivering absolute consistency from press-to-press, shift-to-shift, and plant-to-plant.

A wide range of on-screen dialogue boxes can be viewed in the system, including set-up screens, trends (see Figure 6.7.), LAB view, inspection (Figure 6.8), etc.

With human color experts, according to QuadTech, three to five color corrections on start-up are typical for a new job – or even six to eight for difficult colors. ColorTrack can achieve more accurate color in one or two corrections. ColorTrack claims to be the only software that offers this level of press-side connection between color measurement and ink management.

QuadTech's Color Measurement with DeltaCam makes advanced, in-line spectral measurement affordable. For around the cost of a color register system, printers can utilize accurate, automated L*a*b* measurement on film, paper or board – ensuring that all printed product is within their customers' color specifications.

Printers can reduce time and waste while maintaining color throughout the roll – without the need to wait for a roll change to measure with a handheld device. Operators can spot problems early and make corrections quickly, thus minimizing waste, customer complaints, and rebates.

Automatic and intelligent 100 per cent inspection solutions including the detection of defects such as streaks, hickeys, registration errors, unremoved

matrix, missing labels, in-line absolute color measurement to ISO standards using the GV-Spectro in-line spectrophotometer - based system (Figure 6.10), and barcode verification are all now possible using the Guardian PQV flagship system (Figure 6.11) from **PC Industries**.

The system also offers automatic job changeover on the fly for digital printers to seamlessly continue 100 per cent inspection on press. In addition, automated remote job set-up off press help flexo printers reduce downtime when frequently changing jobs.

PC Industries also has a customizable system that verifies 1D and 2D barcode quality against ANSI/ISO standards. Their compliance package provides a secure

Figure 6.9 Illustration shows the QuadTech SpectralCam

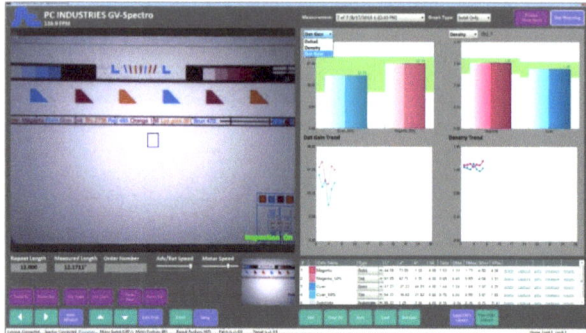

Figure 6.10 PC Industries GV-Spectro in-line spectrophotometer-based system

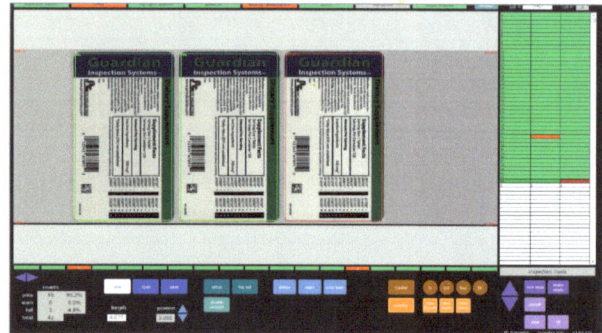

Figure 6.11 The Guardian flagship system for 100% default detection. Source: PC Industries

login, reporting and documentation to allow pharmaceutical printers to fully comply with FDA CFR Part II Compliance. Grading and verification can now be done at higher speeds while maintaining accuracy and minimum code size. The system also can verify human readable characters using OCR/OCV technology.

Another software company that has introduced a new software program for label printers is **One Vision.** Called DigiLabel, it is a comprehensive 2-in-1 system of automated label production which eases complex label workflows for digital label printers and combines print data optimization and label production planning. The benefits of the system include improved transparency, a safe production process and significant improvements in efficiency.

The software includes workflow solutions that enable the entire production workflow to be integrated, standardized and substantially automated. It also provides detailed quality control and optimization of print files and images using RIP and Inksave software, as well as individual solutions for the imposition of PDF files or flattening of transparencies – all done automatically.

DigiLabel makes use of this expertise in the entire production workflow: print data are automatically subjected to a quality control and then optimized. The automated production of die-cutting forms (see Figure 6.12) or laser-cutting configurations, standard cut lines, a white background and the automatic dispatch of a release PDF to the customer for approval saves labour time and staff resources.

The ColorCert Suite of software products from **X-Rite** provides real-time color and print quality process control and reporting to help packaging printers and converters better manage the complexities of CMYK, extended gamut, and spot color workflows, regardless of the printing process, substrate, or industry standard.

The ColorCert Suite consists of three modules specifically designed to help various members of the packaging workflow manage specifications and monitor quality control from the artwork to premedia to the ink room to the press room. The modules include:

- ColorCert Desktop Tools enable the accurate creation of color specifications and process controls at multiple points throughout the packaging workflow.
- ColorCert ScoreCard Server combines reporting and statistical process control to provide brands, packaging converters and premedia professionals with an easy-to-use dashboard for the monitoring of print quality by supplier, plant, machine, customer, work type, and more.
- ColorCert Repository Server, a secure, online asset management solution for storing, managing and deploying packaging color assets so that the most up-to-date assets are always being used and the packaging workflow is optimized.
- For brand owners, ColorCert provides an overview of supply chain performance to ensure accurate and consistent color across regional or

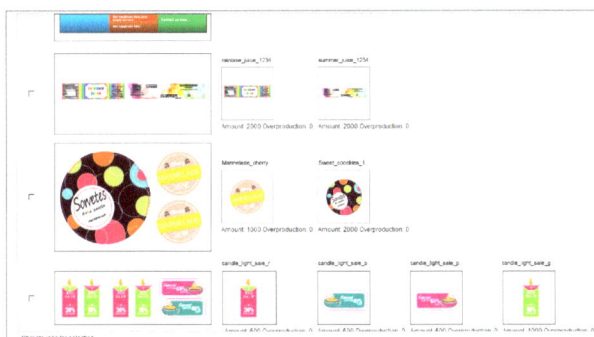

Figure 6.12 DigiLabel saves resources by using cross-customer collect-run production

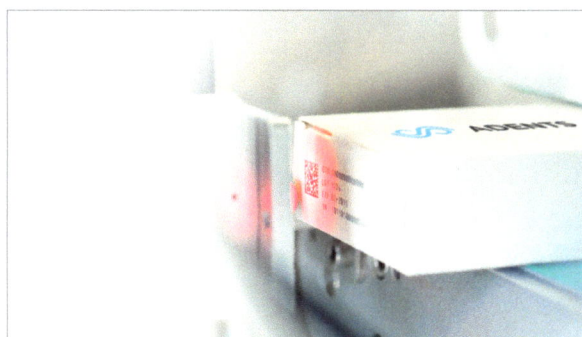

Figure 6.13 Adents Prodigi offers compliance to Level 4 traceability

global partners.

GLOBAL DRUG COMPLIANCE

The demand for unique product identification solutions, fueled by a global increase in regulations in the pharmaceutical industry, where more than 80 per cent of global drug production will need to be compliant with regulations regarding unit identification by 2019, has lead Adents and Microsoft to combine their expertise in unit identification and business intelligence to introduce a new Cloud platform named Adents Prodigi.

This new process, claimed to be the only Level 4 traceability solution (Figure 6.13) that can centrally manage regulatory requirements imposed on the pharmaceutical industry, as well as allow laboratories and CMOs to take advantage of the massive amounts of data generated during the serialization and Track and Trace process, offers extraordinary opportunities in terms of visibility on the life of products, but also represents a major challenge in terms of data management.

Adents Prodigi relies on advanced Microsoft Azure technologies and allows drug manufacturers to securely generate, exchange and control the huge amount of data created through unit identification. The Adents Prodigi platform also integrates data analysis (including Microsoft Power BI) and machine learning tools, providing powerful analytics capabilities for these new sources of information.

In total, Adents Prodigi stands as an ideal solution

to ensure compliance with new regulations while efficiently mining and utilizing new data to better manage production and distribution practices beyond regulatory compliance in terms of:

- Unit identification and traceability
- The fight against counterfeit medicines
- OEE and productivity monitoring across production sites
- Enhanced user experience and personalized communication
- Real-time data visualization and analysis

FINDING THE OPTIMUM SOLUTION

As can be seen from the MIS, software and hardware solutions set out in the preceding paragraphs, there is now a wide range of products that can be used to inspect, check and record almost any aspect of quality control, quality management and compliance. Even the solutions described in this book are far from exhaustive today. More come onto the market every year. Not all are fully integrated with MIS, but the list continues to grow.

Depending on the quality or compliance requirements demanded by customers, label and package printers can now find a solution or solutions for pretty well any quality-related application, market or customer. It is a question of looking to see what is available and test or trial as required so as to find the optimum solution.

Chapter 7

Job costing and shop floor data collection

Once a job has been completed – having passed through all the earlier MIS modules as required from estimating, order processing, production, inventory control – and then finally being shipped, it ideally needs to be costed and an invoice issued as quickly as possible. Additionally, the sales, accounting and management team will want to know if they made a profit on the job. Indeed how much profit?

Costs will therefore need to be allocated against specific customer orders, specific activities or products. It also involves tracking the expenses/costs incurred on a job against revenue produced by the job, comparing quoted costs against actual costs, analyzing variances and ensuring all costs involved in a job have been properly invoiced. This can all be undertaken within the MIS costing module. See Figure 7.1.

Job costing within a MIS system therefore enables the accumulation of the costs of substrates, inks, dies, foils, labor, and overheads for a specific job. Quite simply, Job Costing follows a job through the necessary manufacturing to shipping stages, picking-up the cost of labor, materials, overheads, etc., throughout the entire production workflow. This approach provides an excellent tool for tracing specific costs to individual jobs and examining them to see if the costs can be reduced in later jobs. An alternative use is to see if any excess costs incurred can be billed to the specific customer.

In essence, job costing involves keeping an account of all the direct and indirect costs involved in producing a job and ensuring that:

- All the cost involved in producing the specific job are fully tracked and allocated
- Making sure that all costs are allocated to the customer
- Producing reports that show details of costs and revenue by job.

Management teams will need this information in order to make informed decisions about production levels, pricing, competitive strategy, future investment, and a host of other concerns. Such information is primarily necessary for internal use, or managerial accounting. By tracking and categorizing this information according to a rigorous accounting system, corporate management can determine with a high degree of accuracy the cost per unit of production and other key performance indicators.

Indeed, job costing software within the label or package printing plant today offers easy access to all aspects of company profitability and margins. This can be done on individual jobs, by specific customer, by day, week or month, by machine (conventional, digital, hybrid), by employee if required and by product type. Custom report libraries and custom report writers in MIS now offer almost any variation required.

Quite simply, Job Costing MIS will significantly

change the way that a label or package printing business is run and with the entire data collection, costing cycle and analysis now available to the management the business will undoubtedly become more profitable.

Labor and machine hours (with employee clock-in and out of each job, by function), together with the cost of direct materials, will obviously constitute the majority of direct costs in a label or package printing plant, with companies typically tracking the cost of finished raw materials as direct costs. To these costs must be added fixed costs, indirect costs, or overhead costs, together with any variable costs. These elements are shown in Figure 7.2. So let's examine all these cost elements in a little more detail:

FIXED COSTS

Fixed costs are an operating expense of a business that cannot be avoided regardless of the level of production. Fixed costs are usually used in breakeven analysis to determine pricing and the level of production and sales under which a company generates neither profit nor loss. Fixed costs and variable costs form the total cost structure of a company, which plays a crucial role in ensuring its profitability

VARIABLE COSTS

A variable cost is a cost that varies proportionately to changes in the volume of activity or level of output. For example, a sales rep might be compensated with a fixed salary along with a commission or bonus that fluctuates with sales performance. In this scenario, the commission would fall into a variable cost category, whereas the salary is fixed.

DIRECT COSTS

Direct costs are the costs that are readily traceable back to the specific label or printed package job – as can be seen in Figure 7.2. – and which include substrates, inks, cutting dies, embossing or foiling dies, labor hours, machine hours, shipping costs, etc., which are all required to produce the printed and finished job. Direct costs may also often be subdivided into direct material costs and direct labor costs. Direct costs are also referred to as prime costs.

Costs of materials are determined from the prices of all necessary materials that go into producing the printed and finished job, plus any sales tax, shipping, and other related costs. Any unanticipated price changes can of course complicate this otherwise straightforward process.

INDIRECT COSTS

Indirect costs include plant-wide costs such as those resulting from the use of electrical or gas energy, water, telephones, waste disposal, but indirect costs may also include the costs of minor components such as oil and cleaning materials. While all such costs are conceivably traceable to a costed job, the determination of whether to do so depends on the cost-effectiveness with which this can be done. Indirect costs of all kinds are sometimes referred to as overheads, and in this sense prime costs can be distinguished from overhead costs.

As can be seen, there are a number of elements to job costing, and converters may use all, some or none of them. However, if companies want to use a job costing system or module effectively, then they need to:

1. Be able to track all the costs involved in the job
2. Make sure all of the costs are invoiced to the customer
3. Produce reports showing details of costs and revenues by job.

| ESTIMATING | ORDER PROCESSING | PRODUCTION SCHEDULING | INVENTORY CONTROL | QUALITY CONTROL | COSTING | ACCOUNTING |

Figure 7.1 Job costing is the penultimate module within the MIS system

KEY ELEMENTS OF A COSTING SYSTEM			
FIXED COSTS	**VARIABLE COSTS**	**DIRECT COSTS**	**INDIRECT COSTS/OVERHEADS**
E.G.	E.G.	E.G.	E.G.
PURCHASE OF PLANT, EQUIPMENT AND VEHICLES	COMMISSION	SUBSTRATES	ENERGY
INTEREST PAYMENTS	BONUSES	INKS	WATER
ADVERTISING	OVERTIME	CUTTING DIES	TELEPHONE
RENT	PACKAGING	EMBOSSING/FOILING DIES	DEPRECIATION
TAXES	NON-SCHEDULED MAINTENANCE	LABOR HOURS	EXECUTIVE SALARIES
		MACHINE HOURS	WASTE DISPOSAL
		SHIPPING	OFFICE SUPPLIES
			PURCHASE OF PLANT, EQUIPMENT AND VEHICLES

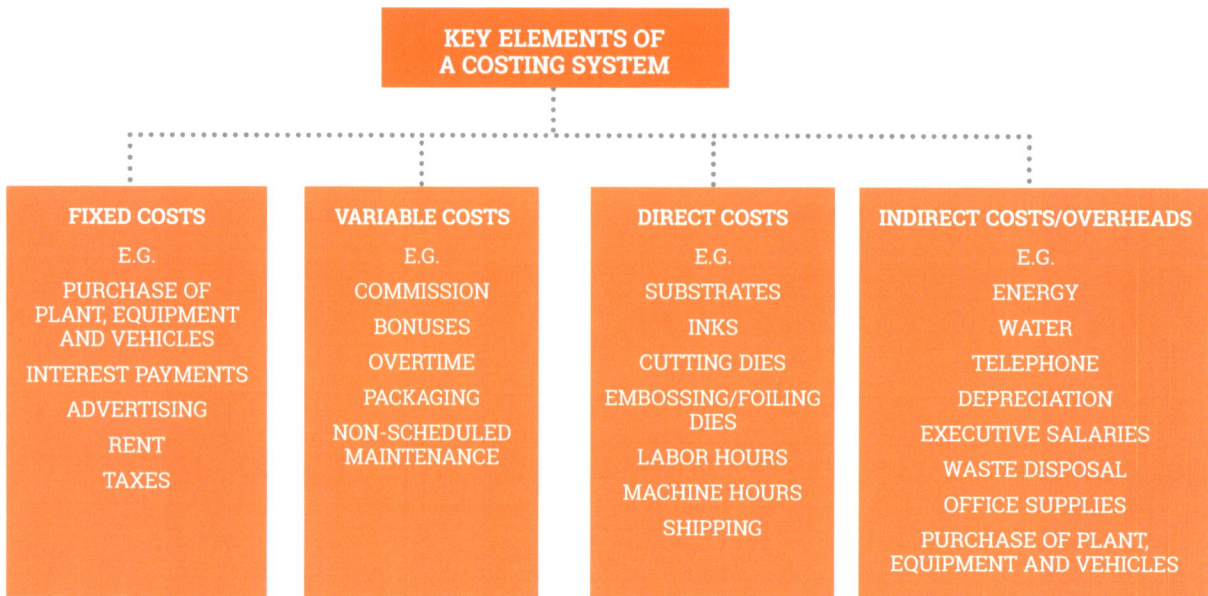

Figure 7.2 Shows some of the key elements in a label or package printing industry job costing module within a MIS

Care should be taken if 'extras' are added to the final costs submitted to the customer. Extras are always regarded with suspicion and customers tend to have a feeling of being 'caught' by what they regard, quite unjustly, as being tricked into paying more than they expected. Ideally the original accepted quotation should quite clearly indicate what is included in the job costs, so that the reason for any extras will be readily understood.

If extras are anticipated before a job is started or even completed, it is advisable to discuss these with the customer to explain why they are arising and maybe give the client the opportunity to review, minimize or cut out the additional item(s). In any event it will help the customer to regard the additional charges as more justified and not something that just appears out of the blue on the job invoice.

ALLOCATION OF MATERIALS IN JOB COSTING

In a label and package printing job costing environment, the materials to be used on a job – substrates, inks, varnishes, dies, foils, etc. – will first enter the plant and are usually stored in a designated warehouse or storage area, after which they are picked from stock and issued to the specific job.

If reel or sheet handling, ink or other waste is created, then any normal amounts are charged to an overhead cost pool for later allocation, while abnormal amounts are probably charged directly to the cost of goods sold. Once work is completed on a job, the cost of the entire job is shifted from work-in-process inventory to finished goods inventory. Then, once the goods have been shipped and are ready for invoicing, the cost of the item is removed from the inventory account and shifted into the cost of goods sold, while the company also records a sale transaction.

Figure 7.3 Scanning of label stock using bar codes and scanners

Label or board stock, inks and other consumable information can today almost always be automatically captured using bar coding and scanners - so providing a comprehensive trail of bottom line materials used information. See Figure 7.3. This information can all then be compared against the original estimate.

Having said this, there are still difficulties encountered in the measuring of ink consumed. This is because ink is sometimes mixed out of other inks without real booking, and secondly, when different jobs are printed one after another with the same ink, it is not easy to divide the total consumption per job. This is why converters may not always book ink. Interfacing with ink dispensing systems, such as GSE, so as to ensure proper booking and traceability of ink, continues to be developed.

Converters that require even more automated shop floor data now have the availability of solutions that provide real-time press room statistics, including the number of labels or sheets produced, material used and running speeds. Data can include full capture of both gross and net labels or cartons as well as length

or quantity converted, so providing accurate cost monitoring control throughout production.

ALLOCATION OF LABOR COSTS

In label and package printing costing or MIS modules, the labor costs may be charged directly to individual jobs, providing the labor is directly traceable to those jobs. All other manufacturing-related labor is recorded in an overhead cost pool and is then allocated to the various open jobs. The first type of labor is called direct labor, and the second type is known as indirect labor. When a job is completed, it is then shifted into a finished goods inventory account.

Once the goods are shipped and ready for invoicing, the cost of the asset is removed from the inventory account and shifted into the cost of goods sold, while the company also records a sale transaction.

Label costing systems can now be linked to the clocking in and out of each employee, with time being automatically recorded to the job docket, or linked directly to shop floor data collection for press running time, number of labels produced, run meterage, etc. (Figure 7.4).

ALLOCATION OF OVERHEADS

With the allocation of overhead costs, non-direct costs will be accumulated into one or more overhead cost pools, from which costs are allocated to current open jobs based upon some measure of cost usage. The key issues when applying overhead costs are to consistently charge the same types of costs to overheads in all reporting periods, and to consistently apply these costs to jobs. Otherwise, it can be extremely difficult for the cost accountant to explain why overhead cost allocations vary from one month to the next.

The accumulation of actual costs into an overhead pool is inherently inaccurate, since the underlying costs cannot be directly associated with a specific job. This means that their allocation to jobs can be a time-consuming process that can interfere with closing the books on a reporting period.

To speed up the process, an alternative is to allocate standard costs that are based on historical costs. These standard costs will never be exactly the same as actual costs, but can be easily calculated and allocated.

The overhead allocation process for standard costs is to use historical cost information to arrive at a standard rate per unit of activity, and then allocate this standard amount to jobs based on their units of activity. The total amount allocated can then be subtracted from the overhead cost pool (which contains actual overhead costs), with any remaining amount disposed of in the overhead cost pool.

The allocation of an overhead cost pool is by definition inherently inaccurate. Consequently, it is best to use the simplest of the above methods to dispose of any residual amounts in the overhead cost pool.

POINTS TO CONSIDER WHEN LOOKING AT COSTS AND COSTING

While the determination and allocation of cost elements mentioned earlier in this chapter may look comparatively simple, there are nevertheless quite a range of things that may need to be considered when setting up an MIS costing system according to Geert Van Damme, Managing Director of Cerm.

Some of the more important of these considerations are set out below.

- The usage of 'fixed' and 'variable' costs can be somewhat misleading. The same words are used to calculate the fixed costs of a job/estimate as the sum of the costs to get the first copy (= set-up) and the variable costs of a job as the production costs per 1000. This is often used to calculate the cost of a range of quantities to be produced, since you only need to do fixed cost + (variable cost * quantity / 1.000).

- In Job costing it is very important to analyze the production. But at the same time, it is equally important to analyze the 'non-productive' time. That is the time in between jobs. The percentage of productive and non-productive time in the total time is essential to estimate how far you can fill up your total capacity with jobs and how well the company did in the current month. After all, the business will only 'win' money with the productive time. What needs to be considered is that calibration, maintenance, start-up, shutdown, errors, breakdown, repair, etc., is just as important to analyze in detail. The shop floor data collection of most MIS-systems will ask

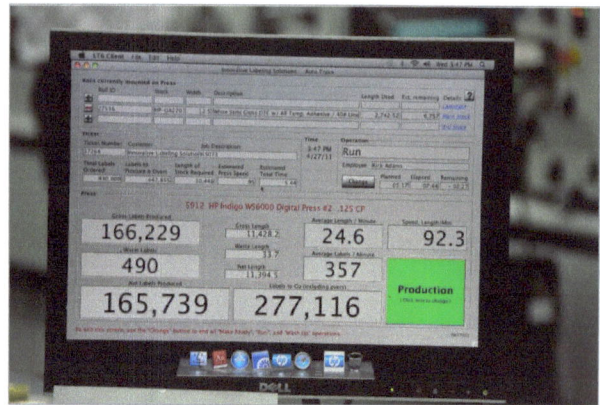

Figure 7.4 Automatic collection of shop floor data.
Source: Label Traxx

operators to 'badge' these non-productive operations and even to comment on the reasons for standstill in order to be analyzed.

- The industry is used to talking about 'direct cost', which is any cost related to the production of the actual job, and about the 'full cost', which is adding the indirect costs (overhead) to the direct costs to end up with the total job cost. The challenge is to find the correct 'parameters' to divide the overhead costs. These costs are to be found not just in the administration, but also in production and warehouse.

- These costs can easily be found, but they cannot be easily divided to the items that will be booked. For example:
  - Should the cost of the warehouse operation be equally divided in $ per 1000 labels (no - since labels can have a different size), or will this be equally divided per 'hour' (again, no - since some operations, like platemaking, do not consume paper). The answer is to find a proper parameter, like $ per m² or a 'handling cost' per paper roll.
  - Or, will the cost of the building be divided equally in $ per hour produced, or should the charge on bigger machines be more than smaller machines?

All of this should result in a model where per 'item' consumed/booked it becomes possible to define 'direct cost' and 'overhead'. This will avoid the company unwillingly making some jobs too expensive because the overheads have been wrongly divided.

And off course, the more that (power - EMS) can be measured effectively, the easier it becomes. But some costs (like the car of the print production manager) will still be divided 'arbitrarily'.

- Another way to look at costing is by judging Added Value. This is the total sales of a job, minus the external costs (paper, ink, plate, outside services). This represents the 'internal value' that is added to the externally purchased things. Using this method it is possible to have a percentage of margin that is equal for two jobs, but as paper is more expensive per square meter for one job than for another, printers sometimes only look at the Added Value that is 'returned' by a job.

  So sometimes the Added Value will be divided by the internal direct costs to have a ratio that will show how many $ will be created for every $ spent in production. A healthy figure would be $1.5. Some printers will even divide Added Value by the number of print hours, since printing is their main business and they have calculated that they should have $250 of 'added value ' per hour printed to have a profitable business.

- There is also something to add about job profitability. Consider when a printer produces for stock and products have to be kept in stock for a long time (e.g. 6 months). This means that the real sales will only be known after 6 months. That's why the production margin of jobs may be judged by comparing the real cost with the inventory value. Then later, a production margin can be shown that will compare the sales with the inventory value. This enables a judgement about jobs to be created immediately after production, even when nothing is invoiced yet (because the job is waiting in stock).

- An interesting element to consider for the future will be the costing of 'ganged' labels that have been produced using a laser cutter. Labels can have a different size and perhaps even a separate finishing path. So how will the job costs be divided 'per label' if the printer wants to make an analysis of the profitability of a group of labels that were produced in various combinations?

- Because it is very difficult to book e.g. ink on jobs, another way must be found to check the estimating model. This can be done by booking the estimated ink cost into the costing system (automatically). This then allow for a more realistic margin per job, but additionally it can be totalled for a longer period (e.g. a year) and be compared with the total volume of purchased ink in the same period.

  This will not provide a perfect solution, but will at least give an evaluation method. The same applies for transport costs. If these are calculated within the per 1000 price, it becomes necessary to book every transport cost onto the correct job, even when the job can be 6 months ago. So here as well, the best solution would be to book the theoretical cost as a real cost and to do a separate analysis for whether the total transport purchase invoices match the total of the booked transport costs over a given period.

- In a further point, it is not easy to divide costs of plates, dies and tools on production jobs. In fact, you sell them the first time and when you need to re-make them, you will not charge them to your customer. So the sale can be long over when costs are still there. So, printers should be careful not to book costs on jobs (e.g. platemaking) when sales will not be on these jobs. The best is probably to do a separate analysis of tooling costs and sales.

ACTIVITY BASED COSTING

Another comprehensive means of determining costs is to use Activity Based Costing in which support activities typically include anything done outside of the normal job production itself and not captured by traditional job costing systems. In other words, what are the real costs incurred in running a label or package printing business?

According to Label Traxx President Ken Meinhardt, 'Many companies use a variety of sophisticated costing programs to determine labor and material costs, but unfortunately, decision makers are forced to 'guesstimate' administrative and support activities associated with a particular job. These methods tend to hide costs of unused capacity, thus giving company managers distorted views of how much excess capacity exists and the associated costs.

By using Activity Based Costing methodology, such as the Litho Traxx's ABM (Activity Based Management) tool, based on Activity Based Costing methods, the focus is placed on the 'support expense' caused by support activities such as estimating a job, entering a job, capturing costs associated with sales activities and customer service, generation of purchase orders, and relevant accounting functions. The tool also monitors utilization of capacity to gain knowledge of 'lost capacity costs' and how these costs affect the overall company health.

A key advantage of the ABM tool is the ability to monitor most activities and collect the information automatically; therefore, relieving the employee of constant recording of their time. This makes solutions like the (Litho Traxx) Costing module an even greater asset. The goal with such a solution goal is to 'help companies to better identify money makers and money losers, find the root cause of problems and correct them, discover opportunities for cost improvement, and develop methods to improve strategic decision making, therefore improving the bottom line.

MANAGING AND CONTROLLING ENERGY COSTS

It is not so many years since power and energy – whether gas, electric or water – were regarded as relatively abundant and not too expensive in the running of a label or package printing plant. Today, these resources are relatively expensive and seem to increase on a regular basis, and can make up a key element of a company's indirect costs, as well as energy usage being a concern in relation to the environment and the use of resources.

Energy consumption is seen as essential in terms of lighting and heating of a plant, in pre-press systems, in the operation of presses – especially in UV ink curing – and during the operation of finishing and inspection.

Financial reports in MIS can help to highlight these

costs. Not unsurprisingly therefore, printers and converters have become ever more aware of the need for energy consumption and cost monitoring, as well as looking at conservation measures. But how can they do this, what measures are necessary, how do they implement these measures and how can they be documented to show their effectiveness. Certainly, energy usage indicators should be quantifiable and measurable, and should relate to all areas of work, as well as being a tool to improve energy efficiency.

Well at least one MIS supplier has addressed this issue. Sistrade has developed an Energy Management System (EMS) that incorporates a tool that enables the control of energy consumption, as well as identifying which cost center causes higher than expected consumption, so making it possible for management to take appropriate actions. Recording of data within the EMS is undertaken in a simple and automatic way, enabling quick and easy analysis of the data and any necessary actions to be taken accordingly.

JOB COST REPORTS

Virtually all MIS costing modules today will be able to provide label and package printing management teams with a detailed job profitability report and job profitability summary analysis in comprehensive reports that show revenues, standard costs, billed costs, the gross profit margin by value and percentage, grouped by customer.

Within each customer report it will show each invoice and each item with the profitability detail by item. These reports may be filtered by customer to see the details for a specific customer, or filtered by margin (see Figure 7.5) to see the most or least profitable items for any time period. Keep in mind that the standard costs you see on these reports are theoretical costs, and are not used in the financial statements.

It is also possible to produce tables or graphs to show materials consumption in square meters and where all types of wastage has occurred (Figure 7.6) as well as comparison tables for calculated and consumed paper, and between calculated and produced products in both meters and square meters (see Figure 7.7).

The job estimates vs. actuals summary and job estimates vs. actuals detailed reports are able to

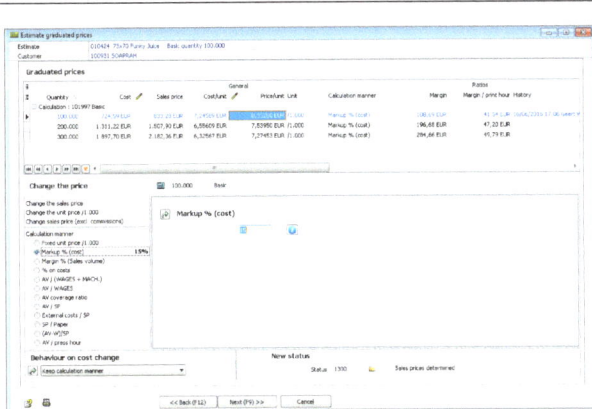

Figure 7.5 Determination of selling price, based upon cost per quantity and margin. Source Cerm

compare quotes to bills and invoices, showing the variance between estimated costs and actual costs and the variance between estimated revenue and actual revenue. These reports are useful for controlling costs during the progress of a job, and as a guideline for preparing quotes for new jobs.

MONITORING OF ADDITIONAL COSTS

An important element of job costing is to look for any unexpected costs – such as artwork amendments or additional proofing or packing charges – which can make a big difference when it comes to job profitability. Some MIS suppliers (such as Tharstern) provide for this by offering an Extra Costs tool which enables the label

or package printing converter to view all costs incurred on each job, thereby enabling them to make informed decisions about whether any extra costs can be passed on to the customer. It can also enable the company to learn lessons for the future.

THE ROLE OF AUTOMATED DATA CAPTURE TECHNOLOGY

Modern MIS technology, real time shop floor data capture and integrated real time production monitoring and production statistics today are able to provide managers with ever more quick and accurate job cost information, making it far easier and simpler to manage jobs with minimal keyboarding and greater accuracy. Much of this is described elsewhere in other chapters in the publication, but can be summarized again as follows:

1. With MIS automated workflow, source documents may only exist in the form of computer records
2. Barcode scanning of incoming materials (and outgoing if required) can be commonly used to enter details of 'goods in' and 'goods out' directly into the computer
3. Other Online information recording and tracking of time, quantity, waste, etc., directly from press or finishing line is becoming increasingly popular
4. Electronic Data Interchange (EDI) technology for ordering materials from suppliers will eliminate or minimize data entry
5. Employee log in can be captured by computer entry, barcode or job docket scanning,

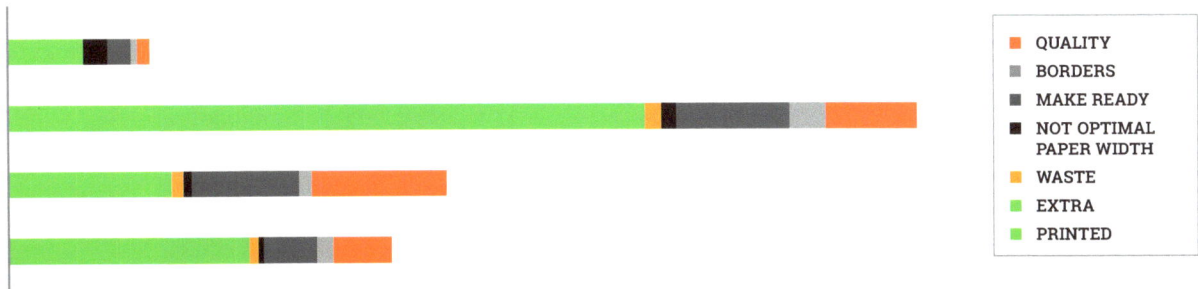

Figure 7.6 The graphs show square meter consumption in good (green) and all types of wastage (colors). Source: Cerm

General	Paper consumption		Production quantities			
Job	Calculated, gross (m)	Consumed (m)	Calculated, net (m)	Produced (m)	consumption	Ordered
963230	942,01	954,00	865,00	847,70	314,82 m²	242,44 m²
963180	1.073,61	899,60	730,00	730,00	314,86 m²	216,08 m²
962633	1.194,19	1.210,20	481,01	481,01	316,06 m²	107,75 m²
964150	1.209,05	1.130,10	913,87	913,87	316,43 m²	227,55 m²
962350	1.030,00	905,10	531,71	531,71	316,79 m²	167,49 m²
963532	1.037,78	1.057,80	545,21	570,13	317,34 m²	135,21 m²
961952	947,02	914,30	525,00	525,00	320,00 m²	165,38 m²
962170	728,97	972,10	484,32	484,32	323,71 m²	149,16 m²
963590	1.002,83	1.087,40	692,00	692,00	326,22 m²	182,69 m²
962637	1.184,29	982,20	724,33	760,55	327,07 m²	219,47 m²
963533	830,31	935,90	476,25	682,63	327,56 m²	153,73 m²

Figure 7.7 Comparison between calculated and consumed paper and between calculated and produced products in meters and square meters. Source: Cerm

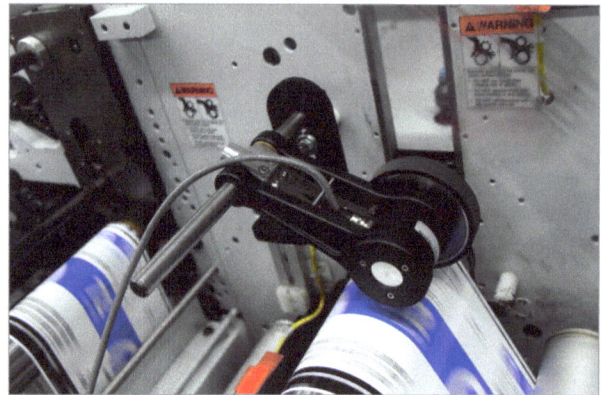

Figure 7.8 EFI's Label Count for up-to-the-minute production and costing data

Certainly, Shop Floor Data Capture today can provide more individual and immediate advantages to the label and package printer, in one hit, than probably any other option. Indeed, automated data capture should no longer be regarded as an option but as an essential part of a fast moving, accurate and effective, modern Management Information System that offers numerous benefits, including:

- Press and ancillary equipment operators no longer having to use their time struggling to complete a manual time sheet every day.
- Accuracy of the data captured, which exceeds anything collated from manual time sheets.
- Elimination of the office time wasted each day in keying in the hand written time sheets.
- The ability to view live on screen the current position of every job, of every data collection point and of every operative (all updated every few seconds). This can be particularly useful for production controller personal and line managers.

The latest MIS and automated systems use environmentally protected, diskless, PC's as terminals, with full-screen color monitoring and a choice of input devices – keyboards, barcode readers, pressure pads, or touch screens. All data is captured live to a central file server so that each individual terminal can be turned off, or temporarily used for other purposes without any loss.

Shop floor production intelligence is also obtainable through solutions such as EFI's award-winning patented shop floor production intelligence platform (Auto Count, see Figure 7.8) that allows printers to automatically collect accurate, up-to-the-minute production data including counts, press status, speed and other critical information directly from equipment in real-time.

Through full plant visibility and data-driven reporting, now in a browser-based environment, label and packaging printers can have all of the information they need to make decisions in seconds.

The use of full screen monitors will also allow for far better error checking of the input; permit viewing of the Job Instructions and allows alterations to be flagged direct to operators. Production managers can use any terminal to gain access to the Planning Board or any other part of the system (with a password) and do not have to return to the office to find the next job due on press and whether the materials are ready.

Using sophisticated MIS and data capture technology within the label and package printing plant now ensures that job costing and invoicing can be carried out with minimal human entry or processing almost as soon as a job has been completed and shipped, with reports available showing all aspects of production and job profitability. For busy plants, such technology is invaluable.

Chapter 8

Accounting and financial management

The Accounts operation is at the heart of any label or package printing industry management information system. It operates the general ledger/nominal ledger where all financial transactions are received, processed and summarized in real-time. The results of these transactions are shown in financial and executive summary reports which provide the management team with financial forecasts, profit and loss data, and all the usual financial management data provided with Ledger Analysis tools.

Using real time data it is possible to view critical business data at a glance, have on-screen up-to-date management reports, track the company's progress against key business performance indicators, obtain an instant picture of how the business is performing and quickly highlight any areas in need of attention.

If required, a whole variety of more comprehensive management reports can be prepared and presented using powerful report writing features that enable users to custom design their specialized reports. These will enable the management team to drill down much further for more detailed analysis: the most or least profitable type of work; opportunities for more profitable jobs; where to target new business.

The Account's department is also involved in the invoicing process, in credit control and surveillance, debt chasing, handling tax rates, pre-payments, possibly working in multi currencies and dealing with accounts receivable.

With the right software solutions the accounts department are also able to manage employee expenses, record and pay supplier invoices, manage cash flows, reconcile accounts and offer historical reporting. Some of the latest MIS versions also feature currency handling solutions for global operations, forex (foreign exchange) sourcing or sales transactions.

Coming at the end of the MIS workflow modules that have been discussed in this book, the accounting and financial management process should be fully integrated and be able to draw on – with highest degree of speed and accuracy – all the information already entered, updated and confirmed as a job passes from estimating, through order processing, production, inventory control, quality control and costing. This workflow process can again be seen in Figure 8.1.

The benefits of a good Accounting and Financial Management System are numerous and can incorporate multiple functions that include:

- Daily sales and bookings
- Monthly accounts of transactions
- Open and failed quotes (in total and by sales person)
- Accounts receivable
- Accounts payable

- Aged balances
- Credit card processing
- Cash Flow and EBITA analysis
- Enhanced payment collection using electronic reminders linked to customers
- Sources and uses of funds
- Ledger and Journal
- Monthly Return and Statement Processing
- Jobs booked by value
- Customer credit limits integrated with order entry and production
- Estimate and credit limit warnings
- Dispatch control against jobs on stop
- Credit notes issues, by reason
- Trial Balance and Balance Sheet
- Payroll Processing
- Taxation
- Annual reports and financial statements
- Auditing
- Debtors
- Bank and Credit Card reconciliation
- Forecasting
- Budget vs Actual review reports
- Multi-currency transactions
- Fixed assets and depreciation tracking
- Expenses handling

THE ROLE OF DASHBOARDS

Some of the MIS financial management systems available today also offer 'Dashboards'. Dashboards make use of color, graphs, symbols, and charts to help users quickly and easily review critical management and financial data. This enables companies to take an immediate pulse of each department and quickly pinpoint problems.

By combining data from throughout the MIS

Figure 8.2 An example of a custom dashboard report looking at label company performance. Source: Label Traxx

system, dashboards create a visual command center where it becomes possible to monitor company performance (Figure 8.2) and launch profitable initiatives. Since getting to the root of an issue may require more than a glance, each dashboard can have the ability to provide multi-level drilldowns, so as to quickly get the company back on track. Each user of these systems can customize their dashboards by selecting from a list of available options that include cash position (Figure 8.3), net profit performance, late jobs, jobs due out each day, delivery by month (Figure 8.4) work-in-process analysis, sales metrics (Figure 8.5) and underperforming employees.

When looking at the cash position and trends, as shown in Figure 8.3, there are a range of graphical visuals displayed that can be used to show elements such as cash receipt trends, cash disbursement trends, days sales in accounts receivable and days sales in accounts payable.

Figure 8.4 is able to offer the management team the ability to analyze on time delivery by month and indicating the target goal (the red line) and the actual delivery performance that has been achieved.

On the left of the screen can be seen a listing of

ESTIMATING → ORDER PROCESSING → PRODUCTION SCHEDULING → INVENTORY CONTROL → QUALITY CONTROL → COSTING → ACCOUNTING →

Figure 8.1 Accounting and financial management comes at the end of the MIS business management workflow

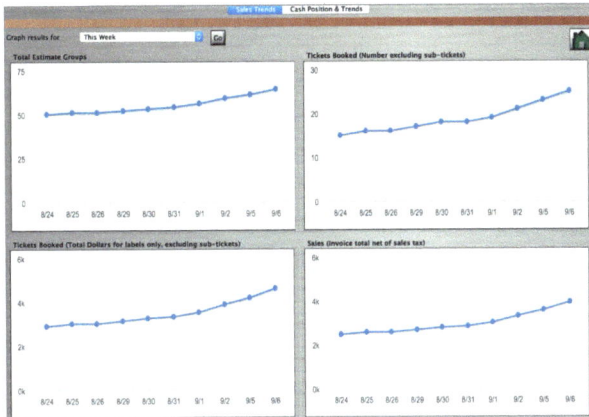

Figure 8.3 Provides an executive report that monitors the cash position and cash trends. Source: Label Traxx

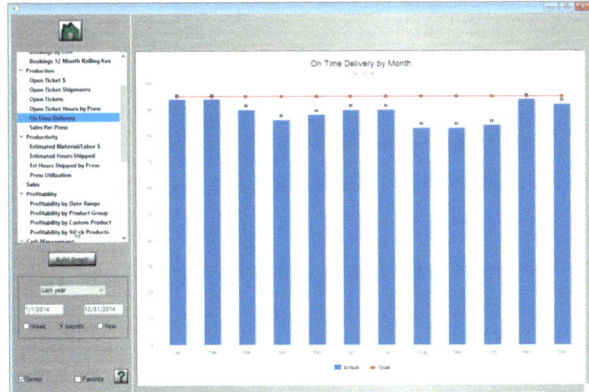

Figure 8.4. Shows on-time delivery by month. Source: Label Traxx

some of the other management dashboard options available for regularly managing the performance of the company as frequently as required – weekly, monthly, yearly or on-demand.

It is not really feasible in a publication such as this to show examples of all the many different levels of sales analysis, bookings, production performance trends, delivery statistics, cash flow against target, and much more.

Suffice to say that a good label or package printing industry financial and accounting MIS will have the capability of analysing and reporting on pretty-well any aspect of the company's performance over any timescale required. It's largely a question of sitting down with the MIS supplier and discussing what exactly is required, in what format and how often.

Businesses today undoubtedly have the opportunity to both analyze and manage their production and financial performance in a way that has never really been possible in the past. It is now possible to see at a glance what jobs make the best or least profit, what sectors are best targeted, where the highest margins can be achieved, what presses offer the best value, which employees offer the best value to the company. In short, there should be no

excuses for poor performance and profitability as everything should be known, analyzed and dissected. Problem, non-performing or loss-making jobs can easily be seen, and hopefully remedied.

Figure 8.5 for example, provides a one week sales trend and cash position analysis. The top left graph shows the trend for cash receipts; the top right graph shows the trend for cash disbursements. Moving to the bottom left, the graph highlights days sales in

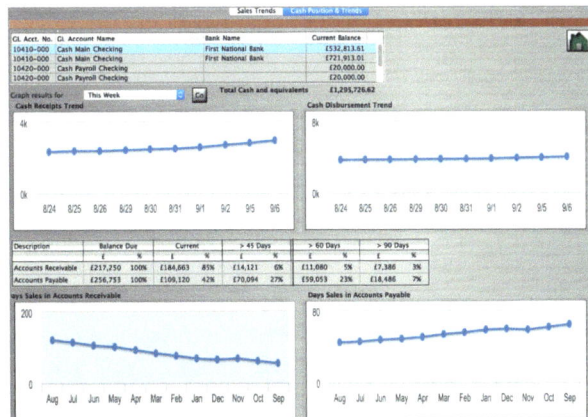

Figure 8.5 Provides a one-week sales trends/cash position analysis. Source: Label Traxx

Figure 8.6 Periodic analysis of sales invoices. Source: Tharstern

Figure 8.7 Job traceability report with production steps (upper part), production, products and material consumption details (lower part) for one job. Source: Cerm

accounts receivable, while the bottom right graph shows days sales in accounts payable.

With some MIS suppliers, 'Accounts Receivable' may be a separate fully integrated, fully functional module for managing billing, cash receipts, and receivable records that is proven to increase cash flows. All elements of creating and accurate invoice is done within the MIS, all job related purchase orders, all shipping and other charges or 'extras' – such as special tooling, handwork, etc. – are brought together for review before the final invoice is posted. The system should also ensure that all shipped items are invoiced or, if required, create multiple invoice types to combine shipments, handle credit notes, or miscellaneous sales.

An important element of financial and accounts software packages is that of credit control. Powerful credit control integration will ideally be checking credit limits against work in process and account balances, as well as checking accounts on 'stop' status. This will ensure that jobs are not commenced in error. There should also be automatic live credit position viewing/checking at estimate, order, production and shipping stages, as well as the capability to flag and post invoices in dispute whilst queries are in existence.

Another features of many accounting and financial packages is the prompting of users when invoices are posted with invalid general or nominal ledger codes, allowing user to correct the code or alternatively direct to a suspense account. Indeed, sales and purchase invoice analysis and reporting should be a regular daily,

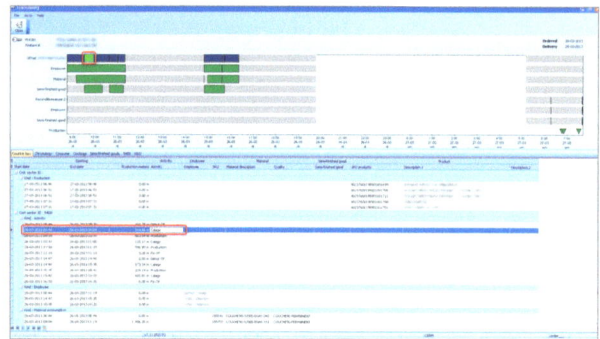

weekly or monthly activity as required (see Figure 8.6).

There can be little doubt that having a single system that manages customers, from the initial order entry through to final invoicing will significantly simply the whole production and administration process. Even if a company already has one of the leading accounts packages, such as Sage 50, Quickbooks Pro, Quickbooks online, or other SQL-based packages, these can usually be fully and seamlessly integrated into most MIS solutions.

If integrated with shop floor data collection some of the MIS management tools will also offer press operators a tool to consult their works schedule per machine and let them indicate what they are doing, and produce job traceability reports showing production steps, production, products and materials consumption details (Figure 8.7). A change in schedule is seen automatically on the shop floor, no need to redistribute lists. Create semi-finished goods to track printed reels. Clockings and consumptions fill the post calculation and gives an insight into the profitability of your finished goods.

Accounts Receivable and Accounts Payable provide all the inflows and outflows, while the General Ledger completes and complements the financial structure of the business. Being able to understand how the business is doing at any time is all about knowing and understanding the production and financial information and numbers. Having this information in real time, from a single source, whenever it is needed, is key to running a financially successful label or package printing business.

Quite simply, today's Accounting and Financial Management MIS software modules should enable label and package printing companies to operate with all the financial information they need to run a profitable business at their fingertips – almost instantly. Businesses should not really be getting into financial problems if they analyze the financials regularly and take appropriate remedial action(s) as soon as problems are identified.

Sadly however, studies from the UK, India and elsewhere tend to indicate that perhaps a quarter or third of label converters have a financial position that puts them into the caution or danger area in terms of financial viability – often on a quite long-term basis. Improved financial management and control using good accounting and financial management MIs software will hopefully start to reduce these kinds of issues in the future.

Chapter 9

Workflow automation – today and tomorrow

The label, and more rather more recently, the package printing sectors, have been undergoing some quite fundamental changes in recent years. The introduction of digital printing alone has brought significant change to the way short printing runs are produced, with sophisticated digital front ends, quick job changeovers, the opportunity to easily run multi-versions and variations, the ability to offer personalization or sequential coding and numbering, and much more.

Not surprisingly, the conventional analogue press manufactures have been coming up with their own innovations to reduce set-up and down time using servo drive technology, quick cylinder and die changes, press digitization, the introduction of job data/definition files (JDF) and shop floor data capture to transfer data to presses and, increasingly working with companies such as Esko and the MIS software companies to move towards ever greater pre-press and workflow automation.

Then of course there has been the introduction of hybrid press technology, bringing analogue and digital print platforms together in one press line and integrating their production, as well as the ongoing growth of quick change slitting and finishing line operations and the use of laser die-cutting.

Add in the developments taking place in integrated production control, cloud computing, WiFi control and robotic handling and the underlying message coming from the most recent label and general printing exhibitions seems to be that the future of label and package printing is undoubtedly going to be very, very, different to that found today.

Already being hinted at during Labelexpo Europe 2015 and followed at Drupa and Labelexpo Americas in 2016, the trend is towards press and finishing line automation, self-managing presses, cloud computing and cloud-based assistants, smart data management and smart printing systems, WiFi control, and even fully hands-free and totally lights-out production, is something that is now being applied across the whole printing press and finishing line community, whether analogue or digital, sheet-fed or web-fed, narrow or wide web, and into all aspects of converting and finishing, from 100% inspection and barcode verification to slitting and die-cutting, cold foiling and spot or gloss varnishing.

No matter whether printers and converters producing labels or printed packaging are looking at printing self-adhesive labels, shrink sleeves, wet-glue labels, flexible packaging, sachets or pouch production, the message from MIS, press, inspection and ancillary suppliers is now pretty-well much the same: responsive and powerful solutions that include full JDF integration across the whole shop floor, ever-more innovative production control tools, the move towards the creation of 'smart factories' and more efficient ways of managing resources. An indication of the automated and MIS integrated streamlined factory of the future can be seen in Figure 9.1.

There seems little doubt that automation of both administration and production continues to be the key to reducing company overhead costs and minimizing or eliminating the bottleneck caused by many small jobs in production. Reducing or eliminating unnecessary operator intervention and increasing the reliability of the data flow can only improve the bottom line of a business.

JDF (Job Definition Format) was created by the printing industry to standardize the information flowing between the management information system and pre-press software or equipment. This standardization provided consistency and reliability from job to job, and has enabled MIS suppliers, including Cerm, Label Traxx and Tharstern to provide a link employing JDF/JMF technology to connect to pre-press software, the first being Esko Automation Engine (formerly BackStage). This bi-directional communication assures that job status is visible throughout the label or package printing company. File planning tools, coupled with Automation Engine step-and-repeat

tools can cut up to 90% of the time from large multi-version orders.

- Such bi-directional communication now provides for immediate status updates when files are received and a proof is ready or approved
- Simplification and structure for the management of art files for each label or pack.
- A complete web-to-print flow with optional human intervention.
- Notification to the customer that the proof is ready for viewing when AE has prepared it and uploaded it to the MIS.

More recently, JDF/JMF technology within digital printing has been extended to connect with AVT inspection and ABG finishing line equipment, enabling the transfer of typical JDF functions such as detailed production parameters, including label dimensions, press, finishing instructions, registration marks, color strategy, step-and-repeat details, and more. Typical JMF functions can then be fed back to the MIS from

THE LABEL PRODUCTION PLANT OF TOMORROW

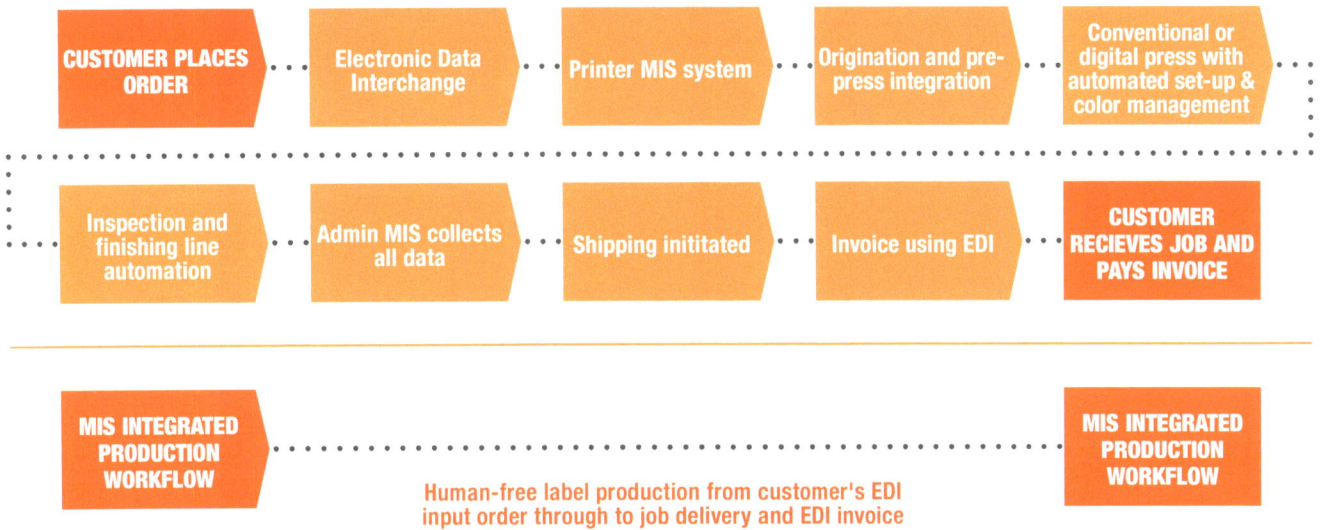

CUSTOMER PLACES ORDER · · · Electronic Data Interchange · · · Printer MIS system · · · Origination and pre-press integration · · · Conventional or digital press with automated set-up & color management

Inspection and finishing line automation · · · Admin MIS collects all data · · · Shipping inititated · · · Invoice using EDI · · · CUSTOMER RECIEVES JOB AND PAYS INVOICE

MIS INTEGRATED PRODUCTION WORKFLOW · · · · · · MIS INTEGRATED PRODUCTION WORKFLOW

Human-free label production from customer's EDI input order through to job delivery and EDI invoice

Figure 9.1 Automate the production process wherever possible and integrate pre-press and production, including integrated die-cutting and 100% vision inspection, with business and administration systems to cope with lots of short to medium run orders and to streamline operations and enhance profitability

the Automation Engine, inspection or finishing equipment. See Figure 9.2 to view typical JDF/JMF functions.

What is also increasingly evident, is that the label and package printing industry is continually seeking ways to further integrate MIS with increasingly automated pre-press, press set-up and selected finishing operations and to remove the risks of human error, as well as being able to handle an ever increasing number of shorter runs and the challenge of facing a lack of skilled operators.

Such moves have undoubtedly been driven by the impressive rise of digital (and hybrid) print for labels, which has extensively pushed both MIS and pre-press software for new solutions. Quite simply, this pressure has been to integrate and simplify every step of label management, pre-press and production.

To remain profitable today, label printers and converters must ensure their pre-press and production workflows are integrated within their business and management operations and connected with their entire supply chain, 24/7, wherever that may be in the world.

It is these developments and innovations in workflow automation that are explored in this chapter, examining some of the evolutions emanating from key industry suppliers and hinting at where the label and package printing industries industry may be in just the next five to ten years.

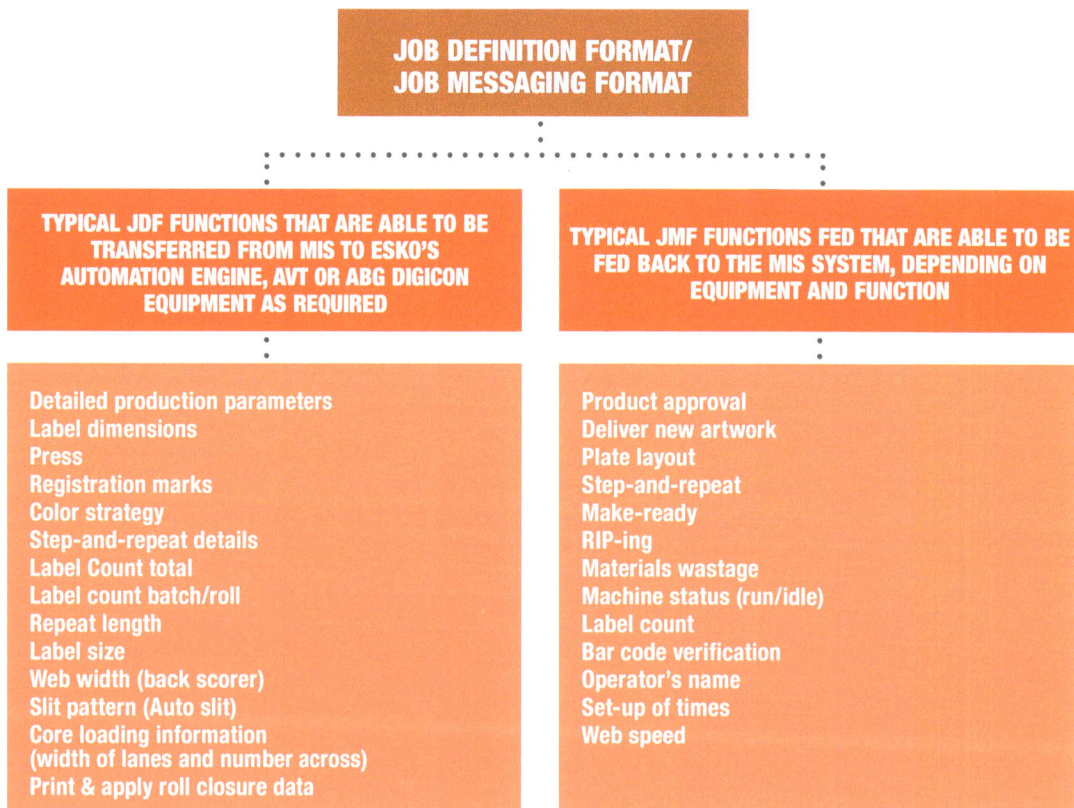

JOB DEFINITION FORMAT/ JOB MESSAGING FORMAT

TYPICAL JDF FUNCTIONS THAT ARE ABLE TO BE TRANSFERRED FROM MIS TO ESKO'S AUTOMATION ENGINE, AVT OR ABG DIGICON EQUIPMENT AS REQUIRED

Detailed production parameters
Label dimensions
Press
Registration marks
Color strategy
Step-and-repeat details
Label Count total
Label count batch/roll
Repeat length
Label size
Web width (back scorer)
Slit pattern (Auto slit)
Core loading information
(width of lanes and number across)
Print & apply roll closure data

TYPICAL JMF FUNCTIONS FED THAT ARE ABLE TO BE FED BACK TO THE MIS SYSTEM, DEPENDING ON EQUIPMENT AND FUNCTION

Product approval
Deliver new artwork
Plate layout
Step-and-repeat
Make-ready
RIP-ing
Materials wastage
Machine status (run/idle)
Label count
Bar code verification
Operator's name
Set-up of times
Web speed

Figure 9.2 Typical JDF/JMF functions that enable bi-directional communication between MIS and pre-press, inspection and finishing

PRE-PRESS AUTOMATION

As already mentioned above, integration of MIS and pre-press automation software (such as Enfocus or Esko's Automation Engine) for example, can now ensure that customer JDF 'job' information from estimating files or job tickets is used to automatically create a new pre-press job and deliver new artwork, make it print-ready and prepare proofs – with status updates on everything from plate layout, RIP-ing and plate making being sent back to MIS, or direct to a digital press.

On output of the production job an XML is created detailing all job components. This collects the submitted job and compares the job properties to the currently supported JDF specification. If the conversion is successfully completed, then a JDF file is submitted to the pre-press or pressroom engine. Additionally, whatever printing process is being used – flexo, offset, letterpress, digital – brand colors can all be controlled, accurately and consistently.

Such workflow automation can now also enable new or reprint jobs to re-use an existing plate set, or an existing cutting die from store, or be used to enable a slitter operator to retrieve previous slitter instructions to automatically re-set the slitting knives.

Other areas of automation co-operation between pre-press and production includes that which is taking place through software and systems integration suppliers such as, Esko, Cerm, Label Traxx, Tharsten, ABG and AVT. Cooperation between Esko and Cerm for example has delivered a seamless integration for product approval and production jobs between the MIS and pre-production environments. Their combined solution ensures the highest levels of efficiency, not only for administration and pre-press functions but also for printing and finishing.

Label Traxx too, have also been developing their MIS system in cooperation with strategic partners including Esko, Rotometrics, HP Indigo, Xeikon, etc., to create a number of tools specifically targeted at streamlining digital workflows, including using JDF to integrate Label Traxx with pre-press to automate step and repeat, generate proofs and other functions.

Certainly, the use of scalable servo drive hardware with intelligent modular design press management software is already increasingly being used to minimize press set-up and make-ready, automate plate or cylinder changes, allow more consistent, repeatable results, and provide converter end-users with greater press flexibility. In-built screens enable all the advanced controls to be visible alongside representative graphics, and provide for easier press commissioning and the simplification of fault finding.

All of this will have an increasing and massive impact on press productivity, freeing up print and label companies to focus on developing their business rather than spending time on managing manufacturing. The use of cloud-based systems, where everything from press performance to planning scheduling is online and instantly available will all be part of the label world of tomorrow.

Pre-press software has certainly made the label printing process much more efficient, as more labels are being ordered through online portals, reducing manual intervention, the time required and the potential for errors. Preflight functionality is seen to be very important for controlling the quality of incoming artwork and this is now highly automated, making it fast and accurate. Automation of trapping, step and repeat, and auto application of marks and bearer bars makes plate preparation of the job much faster and easier, as well.

Customers today undoubtedly require a smoother and more streamlined process from the initial order intake through to production and final delivery. Modern workflows like CloudFlow provide a management 'dashboard' across all production facilities and allow load balancing based on capacity and production needs. Integration of pre-press with MIS and eCommerce platforms now makes it much faster to input orders into production and reduce data entry errors due to re-keying production information.

Esko today provides software to serve the entire workflow from content creation to platemaking. Indeed, as users progress downstream from content creation and get closer to the exposure device (typically the plate imaging system), the more specific software gets to flexo. Esko's Full HD Flexo provides screening algorithms that reproduce more defined highlights and stronger shadows, designed to create the perfect dot on the plate.

For conventional press technology, Esko's Digital

Flexo Suite offers a collection of platemounting software. Automatically and instantly, while a job is sent to the imager, files are created for cutting on a digital cutting table and data files are made for mounting. The flexo plate is cut up into smaller patches to reduce waste, but accurate mounting information is sent to the mounting device. Esko's PreMount workflow is a mounting technique that allows the user to mount flexo plate slugs on a carrier sheet prior to imaging. According to Esko, customers report an average plate wastage reduction of 15% when using the Digital Flexo Suite.

PROOFING PROFILES FOR FLEXO
Apart from Esko's offerings that include proofing, GMG, a developer and global supplier of high-end color management software systems, now offers OpenColor as a method of making proofing profiles for flexo. The company also offers ColorProof, DotProof and FlexoProof as industry software solutions. GMG ProfileEditor contains flexo-specific tools for reliable and precise press-to-proof matching.

Color management is undoubtedly a matter of precision. 'Close enough' is no longer good enough in the critical color world of packaging. It is a reasonable goal to set the flexo printing process to less than 2 Delta E repeatability. The printer's color management systems need to operate within half of that. If the separations and proofs have a variability of over 1 Delta E, then the process cannot achieve the desired goal.

GMG mow offers 12 different color management solutions for flexo and packaging, including PDF image processing and color separation; profile creation for color separations; proofing for flexo; and profile creation for proofing.

ANALOGUE PRESS AUTOMATION
For Heidelberg, now incorporating Gallus, their stated aim is to make printing presses completely self-managing units, where everything from production planning, to consumables ordering, to predictive maintenance, etc., are all generated from the press itself. It calls this its 'Push to Stop' concept, with the presses themselves taking over all aspects of the production.

Heidelberg claim this will have a massive impact on productivity, taking it far higher, and will free up print businesses to focus on developing their business, rather than spending time managing manufacture. Part of the new driverless push is in the Heidelberg cloud based Assistant, where everything from press performance to planning to scheduling is Online and instantly available to management.

Indeed, according to press supplier Comexi, the future is all about printing plants that work 24 hours a day, seven days a week and which can track the work process in one or more plants every day, at any time, from anywhere in the world since it can be managed through a web environment. In their case using Comexi Cloud, a revolutionary software which they claim is the fastest and easiest way to analyze production, which knows and analyzes the incidents time, establishes productive and non-productive meters, controls the execution time and ensures job traceability.

For press manufacturer Nilpeter, they also see the trend in narrow web printing as moving away from seeing the press as a mechanical piece of machinery and towards perceiving it as software-driven, technological equipment. The pre-press job data file (JDF) is rapidly gaining a footing in the industry. The JDF is sent to the printing press. The file contains the job protocol, in CIP3 or similar format, which will transfer job data, such as pressure accuracy, dot gain, register, web control, and cutting depth.

In a similar vein, MPS fundamentally believe that today's market place has an intrinsic need to automate press settings in order to drive down the cost per 1,000 labels, especially for short runs. Therefore, MPS talk about their Automated Print Control (APC) which provides automated servo control of all relevant press settings. Thanks to APC, press settings and controls are extremely easy to operate and replicate through job memory, resulting in virtually no set-up waste for repeat jobs.

From the Bobst Group, who acquired a majority stake in Nuova Gidue, there is also a similar message. Their presses are being equipped with smartCHANGE, a front 'portal' and 4-axis automation system that completely relieves operators of, say, the strenuous exchange of items such as anilox and print sleeves.

DIGITAL PRESS AUTOMATION

Outside of the more conventional analogue presses, albeit ever-more digitized, the message from the digital press manufacturers and digital press software suppliers is also more of the same automation. HP Indigo's WS6800 press for example, delivers high productivity for the vast majority of labels and packaging jobs. Advanced color automation and sophisticated color matching tools make it fast and easy to hit brand colors with extreme accuracy, consistency and repeatability from the first print to the last.

Benefit from the HP Indigo workflow ecosystem including the high-automation Workflow Suite Powered by Esko, and an array of integrated MIS, pre-press and converting solutions from partners. Using HP PrintOS, a secure cloud-based platform that can be opened anytime, anywhere, is a print production operating system with apps that help to get more out of HP Indigo presses, and simplify and automate digital production.

Another company developing automated digital label production is German software manufacturer OneVision, which has introduced DigiLabel, a new software program for label printers designed to improve production processes, provide cost savings and enable a larger throughput.

DigiLabel is a 2-in-1 system of automated label production, and is said to ease complex label workflows for digital label printers. It combines print data optimization and label production planning. The system imports production data and then automatically optimizes it, with the benefits of the system including improved transparency, a safe production process and significant improvements in efficiency.

OneVision is already known as a specialist in the commercial printing industry for pre-flighting and the normalization of print data for error-free printing, offering products that include workflow tools that enable the entire production workflow to be integrated, standardized and substantially automated. It also covers quality control, and optimization of print files and images using RIP and Inksave software, as well as individual tools for the imposition of PDF files or flattening of transparencies, all done automatically.

The company has now brought its expertise to label printing, with DigiLabel (Figure 9.3) making use of this expertise in the entire production workflow: print data is automatically subjected to quality control and then optimized. The automated production of die-cutting molds or laser cutting configurations, standard cut lines, a white background and the automatic dispatch of a release PDF to the customer for approval saves labor time and staff resources. DigiLabel also assembles open orders and handles production planning.

Fully automated, cross-customer collect-run production of labels using the new software is 'unique on the market,' according to the company. DigiLabel assembles open orders and combines labels of various shapes, sizes and print runs to be printed on the same substrate and forwards them to the printers. This reduces production error sources, lowers materials costs by saving on printing and also increases throughput, it is claimed.

Another company working towards full printing process automation is EFI. Working with an AVT solution designed to efficiently support printing process automation and calibration in full synergy with the press's EFI Fiery® digital front end, EFI claim that in addition to enhancing print quality, their dedicated control solutions will increase press productivity through innovative nozzle performance and color control.

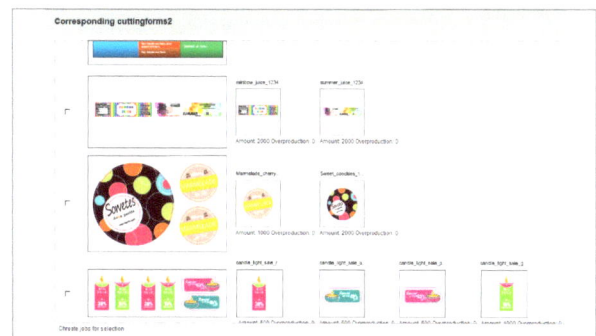

Figure 9.3 DigiLabel assembles orders and combines labels of various shapes, sizes and print runs to be printed on the same substrate

Verification of geometric parameters, such as color-to-color registration, image placement and printing defects detection, are also included. The solution also will monitor data integrity of static and variable content, classify possible print defects and initiate corrective actions.

In addition, EFI's latest ERP-based suite of software (Packaging Suite 4.0) provides packaging enterprises with end-to-end workflows that contribute to profitability by increasing efficiencies. The Packaging suite workflow can be configured to address the specific production management needs for products that labels, shrink sleeves, in-mold labels, flexible packaging, blown film extrusion products and folding cartons.

When dealing with digital printing, EFI explains that the printing method has changed pre-press software. Files must be retained in a format that can be easily changed from digital to flexo and back again for versioning or labeling changes. Native PDF is the best format for this since flattened files are normally no longer editable, and proprietary files must be edited in the original system every time. Native PDF provides complete flexibility. Automation of pre-press has allowed digital printing to reach new levels of productivity and turnaround time by modernizing the pre-press workflow to keep pace with digital output.

SHOP FLOOR DATA CAPTURE

Shop Floor Data Capture today can offer more individual and immediate advantages to the label and package printer, in one hit, than probably any other option. Indeed, it should no longer be regarded as an option but as an essential part of a fast moving, accurate and effective, modern Management Information System that offers numerous benefits, including:

- Press and ancillary equipment operators no longer having to use their time struggling to complete a manual time sheet every day
- Accuracy of the data captured, which exceeds anything collated from manual time sheets
- Elimination of the office time wasted each day in keying in the hand written time sheets
- The ability to view live on screen the current position of every job, of every data collection

point and of every operative (all updated every few seconds). This can be particularly useful for production controller personal and line managers

The latest systems use environmentally protected, diskless, PC's as terminals, with full-screen color monitoring and a choice of input devices – keyboards, barcode readers, pressure pads, or touch screens. All data is captured live to a central file server so that each individual terminal can be turned off, or temporarily used for other purposes without any loss.

Full screen monitors allows for far better error checking of the input; permits viewing of the Job Instructions and allows alterations to be flagged direct to operators. Production managers can use any terminal to gain access to the Planning Board or any other part of the system (with a password) and do not have to return to the office to find the next job due on press and whether the materials are ready.

INSPECTION AND FINISHING LINE AUTOMATION

Integration with camera inspection systems, such as that being undertaken by AVT with their advanced Helios S automatic inspection solution that delivers 100% quality assurance can now enable zero set camera inspection from, say, a Cerm MIS to create the print frames and inspection files per print frame, then print a barcode of every print frame within the job and for every individual 'lane.'

An AVT camera then reads the barcode and verifies the printed output. Electronic interface with camera inspection, using a link to the original PDF for image comparison and instructions for step-and-repeat, can reduce the set-up time of the camera to zero.

Helios S is an automatic inspection system has a user-friendly design that deploys dedicated, advanced algorithms designed to detect any type of defect including color mis-register, color variations, misprints, text errors, spots, splashes, die-cut problems, barcode problems and missing labels. The system works seamlessly on any substrate including self-adhesive labels, thick embossed metalized substrates, highly-reflective holographic foils and laminates.

An optional add-on module for Helios S, with WorkFlow Links to Uniprint, utilizes information recorded on the press to integrate with slitter-rewinders to automatically stop a rewinder, thereby avoiding unnecessary stops on non-defective products and significantly improving overall production efficiency.

In terms of slitter set-up, the setting of slitting knives is undoubtedly one of the most time-consuming jobs carried out on a slitter rewinder. Now however, ABG International have introduced an AUTOSLIT system with auto label gap sensor. The operator simply presses a button, a scanner passes across the web identifying where the gaps are and automatically positions the slitting blades, saving hours of make-ready times each week on short run jobs.

Automatic knife positioning, this time controlled by WiFi and which is able to position a full set of shear cut knives within seconds, has also been introduced by Grafotronic. In their case, every top and bottom knife has an independent drive unit, enabling the operator to adjust single knives if needed.

AUTOMATION OF DIGITAL EMBELLISHING

In a recent development Fusion Technology have announced a new concept with Xeikon that combines full color production printing with digital embellishment of labels and packaging in a single, one-pass and fully digital production process. Over time, Fusion say that this will consist of a series of embellishment modules that are not just put in-line with the press, but are components of an entirely new modular system with the digital front end taking care of the pre-press, data processing, color management and press operation as well as full control and operation of all embellishment modules without manual intervention.

The aim of such developments is to bring closer the reality of fully automated, seamless, unattended, end-to-end production of label and packaging production from customer order to deliver, as already outlined in Figure 9.1. Potential digital embellishment modules include hot/cold foiling, screen prints, matte, gloss and structured flood and spot varnish and a digital braille module.

With Fusion Technology, a print job containing multiple channels defining each aspect of production is dropped into a hot folder and RIP'ed, after which the different channels are sent to the relevant modules including the press – and this without manual intervention. Because every embellishment module is digital, every single design element can be made variable or personalized, which opens up enormous opportunities for new applications. Depending on the requirements, these digital embellishment modules can be positioned before and/or after a Xeikon digital five-color press. The resulting configuration produces a digitally printed and embellished label or package in one single pass.

The Xeikon X-800, its in-house developed digital front-end, ensures a seamlessly automated digital printing workflow while enabling integration with existing workflows and any market-leading third party applications, such as design packages, web-to-print applications and MIS.

INK FORMULATION SOFTWARE TOOLS

Apart from press and finishing line workflow automation there are other areas of label and package printing that can benefit from software tools, such as ink formulation. Here, X-Rite has introduced its updated InkFormulation software tool (Figure 9.4) for providing professional color formulation to printers, converters and ink manufacturers.

This tool offers improved integration so that ink professionals can quickly compare their formulations to a digital color reference for ink color recipe creation, storage, approval and retrieval for offset, flexo, gravure or screen inks.

Ink formulation issues on press can now be resolved more quickly for flexographic and gravure printers as, when press-side color measurements do not meet tolerances, and quality control software reports that tolerances cannot be met with the existing ink formulation, reformulation data is immediately sent to InkFormulation. The ink room can make appropriate adjustments and dispense the new formula, speeding up the correction process on press. This minimizes press downtime and keeps print quality high without the need for the press

Figure 9.4 X-Rite's Ink Formulation software tool

operator to be an ink expert.

The latest InkFormulation update also features enhanced management information system (MIS) integration. This simplifies data interchange with MIS systems whereby the MIS system can request from InkFormulation a new material number based on specifications and naming conventions defined in the MIS, or by referencing a multi-color CxF file stored in PantoneLIVE or elsewhere. This request then appears in the ink kitchen as a job with a target color requiring an ink recipe to be defined. Once the recipe is defined, InkFormulation returns the recipe to the MIS with a bill of materials – a list of ingredients and the percentage for each.

MIS integration helps to close the loop in a production workflow for everything from estimating to invoicing, making it easier to estimate ink consumption by job as well as to reuse existing ink formulations without the need to send a request to the ink kitchen. In addition, all ink recipes required for a given job can be linked to that job, streamlining the workflow even further. This will add significant value at many converter and ink sites.

Additionally, InkFormulation v6.3 allows for the formulation of more than one recipe in the system. The new formulation tab enables formulation and correction of whole jobs coming from quality control software. This new capability will use tags to associate ink recipes to individual jobs.

ROBOTICS

Quite a number of the leading press suppliers are now also readily talking about the greater use of robotics for loading and unloading reels, for bringing reels, inks or cylinders from warehouse storage or pre-production to the press and then after printing taking completed jobs onto finishing or despatch operations, as well as using robotics for on-press handling operations.

It should be noted however, that robotics will ask for better identification of the objects that are being 'manipulated'. So more digital identification units (small printers) will appear, as well as (bar)code-readers. The instructions will be taken out of the computer, at the moment the object's code is read. There will no longer be a JDF 'sending', but a JDF 'retrieval' when needed. This is already the case for example, with the Cerm-ABG-slitter. ABG operators 'scan' the job-code and the instructions are exchanged between Cerm-MIS and the slitter at that moment.

This also applies for AVT camera inspection: the barcode on the digital printed frame indicates the frame-number, and this is related to the 'instruction package' created by ESKO and already prepared in AVT's system memory. A further example of this is a code used for digital laser cutting that will link to the exact CAD-instructions. So the crucial role of MIS will be extended to be the identification-source of all production-components.

ONGOING DEVELOPMENTS IN WORKFLOW AUTOMATION

There are other domains that are typically not covered by MIS software. An example would be dedicated software for machine maintenance programs and/or for spare parts management. Erhardt+Leimer for example has developed a web-based management interface for commissioning, operating and servicing complete web guiding systems via an Internet browser.

All of the necessary devices in the network, i.e. sensors, controllers and actuators, are now networked via Ethernet. No external tools or operating units are required any more for commissioning and operation. Instead, all that is needed is an internet browser on a smartphone, tablet PC or other terminal. This means that operators now have wireless access from any location to the graphical user interface of the device in question, so they are no longer tied to the physical location of the machine.

Production management and quality assurance can now access all of the required information at any time and from anywhere in the world via a web browser, allowing them to view production data quickly and easily in real-time.

Increasingly, MIS suppliers are also being asked to provide an interface so that electronic instructions can be sent to transport carriers, such as DHL and UPS, to manage the transport and shipping of finished goods. Cloud-based systems are also being used to automatically save every inbound and outbound e-mail against customer contacts and produce a sorted trail of previous discussions – all available in the right place when needed.

While recent Labelexpo and print shows have given more than a flavour of where pre-press, press and finishing automation – all integrated with ever-more sophisticated MIS – is now rapidly moving, the next few years will undoubtedly see ever more consolidation and cooperation between hardware and software suppliers.

By then Cloud computing, WiFi control, MIS advances and press digitization shown over the past year or so will all have moved on apace. Indeed, individuals and businesses are already increasingly accessing computing services such as servers, databases, software and storage that are provided over the internet using cloud-based systems. This growth in Cloud computing will undoubtedly bring particular benefits to smaller label converters who tend not to have sophisticated IT and management information systems of their own.

Using the Cloud, label and package printing companies can access any manor of software solutions that will enhance their business without any need to purchase the necessary software or upgrades, and a world where everyone is working from the same software version, without high IT costs. For a monthly fee, or even on a job-by-job basis, the smaller converter can have all of the benefits afforded by larger printing operations.

There seems little doubt that all these developments will bring increasing benefits to both small and large converters over the next few years. Some will already be moving into fully automated, human-free and robotic production; others will be introducing MIS integrated solutions. No matter the level of implementation, the label industry of tomorrow will most certainly be very different from today.

INTEGRATION WITH CUSTOMER PRODUCTION PLANTS – THE ROLE OF VMI

When looking at the future of MIS, workflow automation and integrated production, the industry also needs to understand and assess the future requirements of the production plants of their end-user customers, whether pre-packers or brand owners. More and more is heard about 'vendor managed inventory' (VMI), a streamlined approach to inventory management and order fulfillment which involves collaboration between suppliers and their customers and which changes the traditional order/supply process.

The end aim of VMI is to align business objectives and streamline the supply chain operations for both suppliers and their customers, so providing an improved service, enhanced inventory turn and increased sales. This will mean that the label or package printer will need to fill the warehouse of the customer, based upon the available stock at the customer, and the production schedule and specific requirements of the customer.

This will be true for both printers and their customers, as well as for substrate and ink suppliers and the stock held at the printer. Related to this will be agreements where the customer will only be charged for what he has consumed. So new systems for inventory and invoicing are becoming increasingly necessary.

This will eventually go even further, with the printer delivering direct to the production, packaging, labeling lines of the customer, a few times per day, pallets of labels, flexibles or cartons, nicely ordered with the first scheduled production on top of the pallet – so maybe no warehouse at all in the future.

WORKING TOWARDS
THE SELF-MANAGING PRESS

What now undoubtedly seems to be the aim of most of the leading press manufacturers, is working towards the day when printing presses become completely self-managing units, where everything from production planning, to consumables ordering, to predictive maintenance, etc., are all generated from the press itself, or though integration with MIS systems. A guide to how this is all coming together can be seen in Figure 9.5. This will have a massive impact on press productivity in the future, taking it far higher, and will free-up print companies to focus on developing their business, rather than spending time managing manufacture.

Indeed, the future will increasingly all be about printing plants that work 24 hours a day, seven days a week and which can track the work process in one or more plants every day, at any time, from anywhere in the world – and deliver on-demand – since it can all be managed through a web environment. There are even some systems now that offer the ability to take an order Online, accept payments, pre-flight files, correct them, and send directly to the press without any operator intervention.

How long before human-free, fully-automated label and package printing production, perhaps totally controlled through WiFi and robotics, comes to the market? Probably not too long.

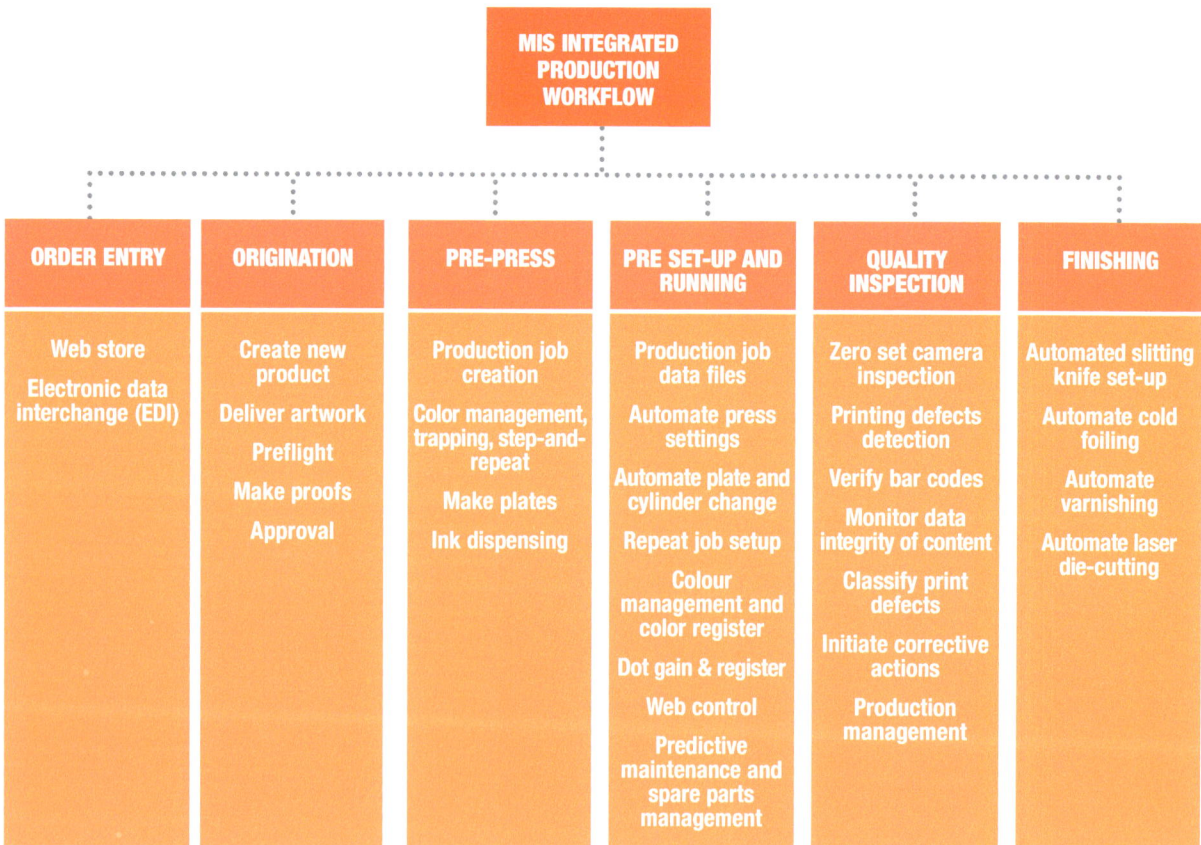

MIS INTEGRATED PRODUCTION WORKFLOW					
ORDER ENTRY	**ORIGINATION**	**PRE-PRESS**	**PRE SET-UP AND RUNNING**	**QUALITY INSPECTION**	**FINISHING**
Web store Electronic data interchange (EDI)	Create new product Deliver artwork Preflight Make proofs Approval	Production job creation Color management, trapping, step-and-repeat Make plates Ink dispensing	Production job data files Automate press settings Automate plate and cylinder change Repeat job setup Colour management and color register Dot gain & register Web control Predictive maintenance and spare parts management	Zero set camera inspection Printing defects detection Verify bar codes Monitor data integrity of content Classify print defects Initiate corrective actions Production management	Automated slitting knife set-up Automate cold foiling Automate varnishing Automate laser die-cutting

Figure 9.5 Working towards full production workflow and the self-managing press

Chapter 10

Choosing an MIS system and finding a supplier

Throughout this book the aim has been to look at management information systems and the software modules that are most commonly available within them, at what the modules offer and how they link together to create sophisticated automated workflows covering everything from administration and business management to production automation.

While there a certainly a handful or so of MIS suppliers that offer dedicated label and/or package printing software – such as Cerm, CRC Information Systems, Edigit, EFI Radius, HYBRID Software, Imprint-MIS, Label Traxx, Tharstern, Theurer and Sistrade – there are probably at least twice that number that offer systems for the more general printing and graphic arts sectors, business forms, stationery, wide format, mail and fulfilment.

In addition there are now other suppliers that offer software packages that can be added on to existing systems for client relationship management (CRM), quality control, environmental performance, warehouse management, shipping, color management, inspection, pre-press, third party connection, accounting, e-Commerce, EDI and other related areas. Indeed the list of add-on software solutions continues to grow all the time.

Many label and package printing companies may already have standard software packages, such as Sage for their accounting and financial control, before deciding to invest in a more sophisticated MIS system to help them administer and manage their,

increasingly sophisticated, administration and production operations. This is particularly true for those that have invested in digital printing and have found they need to handle more and more shorter runs, all kinds of versions and variations, personalization and color management, more invoicing and increased management functions – all of which can cause bottlenecks.

Choosing a Management Information System can therefore be a challenge to any label or package printer that has not really had to think too much in the past about the way they manage their business and may have got by using home grown spreadsheets or custom products, yet now want to provide an increased level of service, reduce turn-around and production times, and manage and analyze a growing volume of sales, production, management and business data.

Few will know initially what they are looking for and what to include in their MIS investment, or really know who to talk to about the challenges and options available. Certainly there will be some soul searching and internal discussions well in advance of actually purchasing a system. Is the company ready for change? Will the staff embrace the change? If not, then it may be advisable to halt the investment process at this stage. Hopefully however, from reading through this book it will at least make it easier for the company to determine what is available, what it is they want to achieve and to have a good idea of how to go about it. However, let's take this a little further and try and set out some guidelines on how to go about successful MIS investment.

CHOOSING AN MIS SUPPLIER

There can be little doubt that chosing the right MIS supplier is an important decision for any label or package printer, probably alongside decisions like buying a new press or pre-press hardware and software. It is important for the future management and profitability of the business, and has implications regarding employees and training.

It is also important to understand the benefits that a universal industry specific MIS can bring to a company, as opposed to a having a number of different systems. It will undoubtedly bring particular benefits for large businesses with a lot of equipment and processes, but even some of the smaller label and package printers – particularly those that have, or are about to invest – in digital printing.

For those in a label or package printing business that are responsible for sourcing an MIS there will be a need to think carefully about what they want to achieve within a complete business-wide integrated MIS implementation, especially on deciding where to best focus attention and on what areas of the business most require help or support to improve efficiency or performance.

Initially, a wish list of features – defined and written down – will need to be prepared as a starting point for talking with MIS and other specialized industry-related suppliers. This list needs to be broken down into categories (estimating, order processing, production management, inventory control, quality control, costing and accounting) in a way that can be given to potential suppliers. The workflow chart (reproduced again below as Figure 10.1) and used in each chapter throughout the book is perhaps an ideal place to start.

It might also help to re-look at Figure 1.2 in Chapter 1 which provides a list of what can be included in a management information systems in terms of Input, Processing and Output. Knowing what the company would like under each heading or category, or more precisiely, what they actually need, is therefore a prerequisit. Finding the right solution and supplier can be a challenging and serious business, but the company must clearly define its requirements under each heading before starting the search

Undoubtedly the most successful installations will be the result of having clearly set-out the company's requirements and then both sides agreeing on exactly what is being supplied, when and at what cost. However, do not make the mistake of having too big a team to draw up the specifications; everyone will be asking for different (often conflicting) requirements and this can end up delaying implementation and maybe an over-specified and much more costly installation. Perhaps better to concentrate on the basics, but which can be readily upgradable.

Having now mentioned 'cost' it should be stated that this is a key factor that needs to be agreed at a

Figure 10.1 The basic functions of MIS and Workflow Automation and how they relate to each other in a typical MIS workflow system

very early stage, even before approaching a supplier. The company should certainly have a pretty clear idea of a budget target or cost limitation before talking to any vendors; something that potential suppliers will certainly want to know.

Key people in the company obviously need to be involved in decision making, with the leadership staying involved, rather than looking to delegate the whole project. Also make sure to use the vendor's experience, use their training services as often as is required, and make sure that the project team meets on a regular basis.

A key point to consider is that with all information management systems it needs to be agreed prior to establishing and implementing that all information assets are corporate assets; that information should be made available and shared (although not all information is available to everyone in the company); and that all information that the company needs to keep and archive is retained and managed corporately.

This can be increasingly important today when internetworked information systems play a vital role in the business success of an enterprise. The Internet/intranet can provide the information infrastructure a business needs for efficient operations, effective management, and competitive advantage.

It perhaps also needs mentioning that is important to choose an MIS supplier that understands the company's specific industry sector needs (labels, cartons, sleeves, flexible packaging, etc.) and that will support the company's business through ongoing growth and other changes. Certainly a close relationship with the MIS supplier will be a vital component in the success of the MIS installation and in the ongoing development of the business.

Probably one of the most common and really expensive mistake company's make when choosing a MIS solution is when the label or package printer gets sold a system by a supplier that doesn't really understand the intricacies (tooling, range of substrates, foils, different printing processes, etc.) of their specific business. There are certainly horror stories about label converters and package printers that have purchased generic print MIS solutions that were built for general commercial printers and didn't fulfill the company's requirements. Implementing a

new MIS is already a major exercise without having to be faced with teaching the supplier about things such as die libraries, unwind directions, foiling and flexography.

Many of the leading label and package printing industry suppliers will advise potential MIS buyers to go and visit some of their MIS customers and discuss the experiences that they have had with sourcing and installing a system before going ahead. Also ask about the quality of vendor support services: maybe they have a 'User Group,' and is the vendor prepared to partner with other suppliers used by the business in, say inspection, pre-press, finishing? They may well also advise testing (or demo 'tasting') the software before purchasing.

Initial MIS software demonstrations today are frequently carried out Online, which makes them both efficient and cost-effective. More than one general demo may be required before moving to a more detailed and specific demonstration using the printer's actual shop floor data – which is really the only way to see how the system would handle the work required.

Members of Trade Associations can also talk to other members and to their peers, as well as read the trade press and visit trade shows such as Labelexpo (an ideal way of talking to and reviewing many different suppliers in a relatively short period of time).

Remember, the overall aim with a good management information system should be to end up with a single system that is able to manage the entire business, streamline the administration process and reduce costs, eliminate errors from the re-entry of data and minimize personnel. It should also be label industry specific.

To repeat from Chapter 1, the key benefits that can be achieved with a good MIS include:
- The ability to identify the company's strengths and weaknesses through reports, sales and production records, etc., and so enable the company to make improvements
- The ability to provide an overall picture of the company and the way it operates
- The opportunity to improve decision making and speed up actions
- The capability to better manage customer information and target sales, marketing and

promotional activities
- The ability to gain a competitive advantage when compared with other label converters

However, even with a dedicated industry specific MIS supplier, it still may need to be decided whether a 'straight out of the box' MIS will provide the answers for everything that needs to be achieved, or whether modifications, additions or basic tweaks will be needed so as to get the most out of the system. Sometimes a bespoke system configured to a company's specific needs maybe a better solution, albeit more expensive. But don't necessarily believe the sales pitch; the label or package printer knows the business best. What's needed is a software supplier that knows about the demands and requirements of the label converting or packaging printing business. But try to focus initially on the basics. Complex systems are not always better, and remember too that software and workflow automation equipment is improving rapidly.

Having said that, the ease with which the MIS is able to integrate with other workflows such as those, for example, used by other market leaders that include Heidelberg/Gallus, HP Indigo, Xeikon, Domino, Esko, AVT, ABG, QuadTech, etc., is becoming increasingly important in providing 'joined-up' workflow as seamless workflow automation continues to develop.

Ensuring that the MIS supplier chosen is easily contactable is also important. Queries need to be dealt with quickly. Ideally, there should be a direct relationship with senior members of the MIS provider so that communication can be undertaken directly with them. Undoubtedly there will be challenges with an MIS roll-out on such a scale, but then the benefits of the investment are almost certainly going to strongly outweigh these. Indeed, why not select MIS (as a tool), based upon the improvements it can make to the profitability and performance of the business: achieving significant improvement in quality and lead time and cost?

INITIAL AND ONGOING TRAINING

One of the most important challenges with investing in an MIS is the people. It is frequently said that people are afraid of change. In reality, people are afraid of loss. Loss of control, loss of responsibility and status tend to resonates through their brains.

Education and training for MIS installation and operation is therefore one of the projects that CEOs, VP's and owners should never look to delegate. Use the investment as a change agent to improve the efficiency, productivity, operation and profitability of the business.

Initially, the MIS team or group will need to meet frequently as a group and discuss ways to improve or modify processes. Meet weekly to begin with, then monthly as the system gets implemented. Involve the software supplier as needed; they have years of experience that the team can take advantage of.

Once the MIS/workflow automation system is up and running it will still probably require modifications to take into account the types of jobs undertaken, new press or finishing equipment being purchased, new people joining the business and a changing customer base. Continuous training at some level will therefore most likely be required.

Equally importantly, with the MIS in operation it will start to generate all kinds of data, reports, dashboards, etc. that can start to be used to make management decisions which, on a daily or pre-determined basis can include:
- Evaluating customer profitability
- Quoting and costing to standards and actual times
- Making process improvement
- Monitoring sales, marketing and customer trends
- Tracking production improvements.
- Examine the procedures in the office
- Reviewing who does what and why? Does it bring value?

It is strongly advised that label and package printers should not pursue investment in digital printing without a completely integrated MIS and pre-press system (such as Esko's Automation Engine) which enables comparison estimating for both digital and conventional presses, eliminates the multiple challenges of being swamped with orders, being even more swamped with artwork, and the requirements of proofing and proof approval.

So what would the summary advice be when looking to invest in and run an MIS?

To begin with:
- Invest early and make it a system team effort
- Look at a complete solution
- Manage the people involved
- Switch to the new system sooner than later
- Leadership MUST be involved

However, such a short list is probably not enough for label or package printers looking to invest in a streamlined MIS for the first time, or for those upgrading from a simple to a more complex MIS system. It has therefore been decided to provide a guideline checklist as an aid for new project managers and project teams.

Hopefully the checklist on the next page will speed-up the steps that need to be taken when looking to buy a system and then provide for a smooth investment and implementation stage of whatever MIS supplier and system has been chosen. If the homework has been well done and the correct software modules have been carefully chosen, then the printer can look forward to a more efficient, streamlined and successful.

If anything, overbuy the system – that is purchasing and then USING as many modules as possible. Automate and leave as little to human error as possible, and pull the plug on the old system as soon as feasible. If it is desired to keep a copy, have this on the oldest and slowest computer and discourage employees from using it.

Once installed and operating successfully, use the data and reports to become a data-driven business. Use the data in the system to improve processes, productivity, systems and procedures and, above all, use the system to ensure the business becomes more competitive and more profitable.

GUIDELINE CHECKLIST FOR INVESTING IN AN MIS SYSTEM

The following checklist is for guidance and is not meant to be fully comprehensive. However, it is hoped that it will be of value to MIS project teams looking to invest in, implement and complete a successful MIS installation. Use as a tick box list if required.

☐ Appoint an MIS project leader and establish a small project team

☐ Ensure the project team meet on a frequent basis

☐ Involve key people in decision making

☐ Determine which areas of the business will mostly benefit from the MIS

☐ Create a wish list of the MIS features that will be required

☐ Decide whether the MIS needs to integrate with other company workflows

☐ Build a clear vision of what needs to be achieved

☐ Establish an investment budget

☐ Research potential MIS suppliers in the media, on websites, at shows, etc

☐ Study websites of 4 or 5 possible MIS suppliers. Ask for information from those that seem to have a good fit

☐ Look for suppliers that best understand the company's specific requirements

☐ Aim for a single system that will manage the entire business

☐ Make sure the potential/proposed supplier is always readily contactable

☐ Visit the potential suppliers' customers and discuss their experiences

☐ Do not schedule any demonstrations until you feel comfortable with the potential supplier

☐ Arrange to have the first one or two demonstrations provided on-line

☐ Then test the software in-house on one of your own jobs

☐ Develop a close relationship with the supplier during installation and implementation

☐ Identify which employees will need to be trained in the use of the new system, and make use of the vendor's training services

☐ Arrange frequent project team meetings during the implementation stages and make use of the suppliers expertise in ironing out any snags and making any modifications required

APPENDIX

Suppliers of MIS and software solutions

The following list of some of the key or most relevant suppliers of MIS and related software to the label and package printing industries is set out in alphabetical order. It is not meant to be definitive list or a recommend choice of supplier, but rather to show what is available. Each company looking to invest will need to undertake its own investigations and research by talking with potential suppliers, undertaking tests or trials, and evaluating findings with employees likely to be involved (and maybe even customers) as to what will provide the best solution.

CERM

Provides management software for label and commercial printers. Cerm is an MIS that totally responds to the specific needs of all narrow web label printers. Based on a specially developed model for this niche market, this software is a great tool to structure the business processes and organize the label printer's administrative workflow.
www.cerm.be

CRC INFORMATION SYSTEMS

CRC is a provider of comprehensive and fully integrated business management products for the label and narrow web industry. Its software has been continuously improved over the past 30 years and is used daily to handle the operations of thousands of users.
www.crcinfosys.com

EDIGIT

EDIGIT International – Enterprise Solution for Printing, Digital and Label Companies. Complete MIS (Estimates, Schedule, Stock/Warehouse management, data-capture, accounting and Business Intelligence). Available in 5 languages.
www.edigit.info

EFI

EFI's Print MIS/ERP solutions provide a solid foundation for your business, collecting, organizing and presenting information in a format that improves communication, speeds production, reduces errors, and boosts throughput. No other provider offers the level of flexibility, integration, automation and smart software offered by EFI. The EFI MIS/ERP solutions are an integral part of EFI's integrated, automated Productivity Suite workflows.
www.efi.comm

EFI PRISM

EFI Prism is a world leading Management Information System (MIS) software and shop-floor management tools specifically for the printing and graphic arts industries.
www.efi.com

EFI RADIUS

EFI Radius (fka 'PECAS Vision') is a fully integrated ERP/MIS system developed to manage the unique business information management needs of packaging and printing companies worldwide.
www.efi.com

ENTERPRISE PRINT MANAGEMENT SOLUTIONS (EPMS)

EPMS offers a completely integrated print management solution with the flexibility and scalability to handle companies of any size that offer digital, sheetfed, flexo, screen, web, large/grand format, business forms printing, and mail and fulfillment.
www.entpms.com

ESKO

Esko customers use many hundreds of different business systems (MIS/ERP). They differ by the printing segment they focus on, in their regional presence and in their support for JDF (the standard language to link them to pre-press workflows). Many Esko customers have also built their own system. Esko prepress workflow can be integrated with business systems (Automation Engine), used for structural design (ArtiosCAD) and as a web-communication tool (WebCenter).
www.esko.com

GLOBALVISION

A leader in quality control technology for some 25 years, Global Vision's complete suite of advanced Quality Control solutions features text-based, pixel-based and Braille inspection technologies that are designed to eliminate printed artwork and copy related errors, providing end-to-end security at every stage of the packaging workflow.
www.globalvisioninc.com

GLOBE-TEK CORPORATION

Globe-Tek is a Canadian Corporation specializing in management information systems for the packaging, folding carton, label, tag and printing industries. Users have the ability to organize production plants and environments without extensive and expensive vendor customization.
www.globe-tekcorp.com

GMG SOFTWARE

This company's DBiz online CRM provides a one stop shop for storing, managing, marketing and analyzing Client Relationship Manager (CRM) data.
www.gmgsoftware.com.au

GMG COLOR

GMG is an award-winning leader in high-end color management and proofing software. With GMG ColorProof contone proofs can be produced to the most diverse international industry standards. GMG DotProof is a solution for cost-effective color-accurate halftone proofs based on the original imagesetter data; GMG FlexoProof is geared towards the flexo and packaging markets.
www.gmgcolor.com

HYBRID SOFTWARE

This company's products have been developed based on input from hundreds of customers in all segments of the printing, label and packaging industry. They provide a unique set of tools for streamlining production operations, eliminating duplicate data entry and human error, and automating production to meet the needs of demanding customers. Products are concentrated in four critical areas: customer-facing web portals, job ticketing and automation, soft proofing and editing, and production workflow.
www.hybridsoftware.com

IMPRINT-MIS

Imprint's MIS for Label production provides the label printer access to fast estimating and order creation with full control of all the production processes through schedule and work job management. When integrating with other Imprint modules such as Raw

Stock and Finished Stock Control, Purchase Order Processing and Sales Invoicing, it provides access to full and detailed information at the click of a button.
www.imprint-mis.co.uk

LABEL TRAXX

Label Traxx from Tailored Solutions is print job management software specifically created for narrow web flexo printers and digital label printers who specialize in all aspects of in-line roll converting, whether hot stamping, rotary letterpress or digital production. The company has continuously evolved with the label printing industry and has customers worldwide.
www.labeltraxx.com

ONEVISION

OneVision Software AG is an international provider of innovative and cost-efficient software solutions for the printing, publishing and media industry. The company's product portfolio ranges from software for prepress to tools for intelligent colour management and image optimization to products for digital publishing.
www.onevison.com

OPTIMUS

Optimus offers a fully integrated solution to a variety of print sectors, including labels, package printing, reel-fed digital, sheet-fed litho, and more, helping customers provide customer services, increasing sales, reducing wastage (time and cost) and delivering complete visibility of a business.
www.optimus2020.com

PC INDUSTRIES

PC Industries is one of the leading suppliers of vison web inspection systems for the printing, converting, packaging, security and pharmaceutical industries and provide digital web viewers, high-speed 100% line scan systems and automatic PDF proof reading systems.
www.pcindustries.com

PRINTMIS

Print MIS software introduces an innovative approach to producing run of the mill quotes, plans, production runs and dispatch schedules with zero tolerance for errors and omissions
www.printmis.com

QUADTECH

QuadTech, Inc. is a world leader in the design and manufacture of control systems that help commercial, newspaper, packaging, and publication gravure printers improve their performance, productivity, and bottom line results. The company offers an extensive array of auxiliary controls, including the best-selling Register Guidance Systems, award-winning Color Control System, and the widely known QuadTech® Autotron line.
www.quadtechworld.com

THARSTERN

The Tharstern MIS is a fully integrated and modular Management Information System that can be configured to meet the requirements of a printing business, with workflow solutions now used in over 700 sites and 7,000 users worldwide. The company's market leading MIS software has been driven and shaped by new technologies, changing trends and user feedback to develop a powerful and comprehensive MIS that can benefit businesses from the onset.
www.tharstern.com

THEURER

MIS/ERP-Software for the printing and packaging industry. C3 offers ready-to go business templates for narrow web label printing, flexible packaging as well as folding carton and display manufacturing enterprises. Integration with Esko Automation Engine for maximum efficiency in pre-press workflow.
www.theurer.co

SHUTTLEWORTH

Shuttleworth is a flexible and configurable print management systems that spans all sectors of the print industry from commercial through labels, packaging, large format to direct mail. The company offer a set of innovative tools including enhanced Dashboards, Customer Specific KPI's, User Definable Reporting, Executive Analysis, Web Based Reporting together with Mobile Apps.

www.shuttleworth-uk.co.uk

SISTRADE

SISTRADE - Software Consulting, S.A. is an information systems company with know-how in software development and consulting service for different activity areas. It provides a range of solution for the label printing industry. It is a completely customizable application, being very flexible to new specifications.

www.sistrade.com

X-RITE

X-Rite is a global company with locations around the world. Experts in blending the art and science of color, the company focus on providing complete end-to-end color management solutions for clients in every industry where color matters. X-Rite Pantone products and services are recognized standards in the printing, packaging, photography, graphic design, video, automotive, paints, plastics, textiles and medical industries.

www.xrite.com/colorcert

Index

www.ingramcontent.com/pod-product-compliance
Lightning Source LLC
Chambersburg PA
CBHW041721210326

41598CB00007B/732